The 4th Degree
A Collection of 123
Miscellaneous Essays

Bryan J. Katz M.D., PharmD
Future Secretary of Education

Copyright © 2024 **Dr. Bryan Katz Publishing**

All rights reserved. No part of this publication may be reproduced, distributed, or transmitted in any form or by any means, including photocopying, recording, or other electronic or mechanical methods, without the prior written permission of the publisher, except in the case of brief quotations embodied in critical reviews and certain other noncommercial uses permitted by copyright law. For permission requests, write to the publisher, addressed "Attention: Book Rights and Permission," at the address below.

Published in the United States of America

ISBN 978-1-960684-23-3 (SC)
ISBN 978-1-961507-85-2 (HC)
ISBN 978-1-960684-22-6 (Ebook)

Dr. Bryan Katz Publishing
222 West 6th Street
Suite 400, San Pedro, CA, 90731
dbjkatz@gmail.com

Ordering Information and Rights Permission:

Quantity sales. Special discounts might be available on quantity purchases by corporations, associations, and others. For details, contact the publisher at the address above.

For Book Rights Adaptation and other Rights Permission. Call us at toll-free 1-888-945-8513 or send us an email at admin@stellarliterary.com

The 4th Degree
A Collection of 123 Miscellaneous Essays
By
Bryan J. Katz M.D., PharmD
Future Secretary of Education

Contents

Astronomy / Cosmology 1
Essay 1: Cosmology - A Brief History 2
Essay 2: Constellations Cassiopeia 6
Essay 3: Identifying Constellations for Amateur Stargazers 8
Essay 4: How to Use Constellations to Tell Time 11
Essay 5: The Constellation Cygnus 13
Essay 6: Edwin Hubble and the Hubble Constant 15
Essay 7: How Space and Time Became Space-Time 18
Essay 8: The Constellation Leo 20
Essay 9: Messier Marathon 21
Essay 10: Orion the Hunter 23
Essay 11: The Sun, Moon, and Tides 25
Essay 12: Ursa Major Notable Features of the Constellation 27
Biology & Ecology 29
Essay 13: Bioluminescent Fungi 30
Essay 14: Copper Toxicity in Sheep 32
Essay 15: Eukaryotes - Plant and Animal Cells 34
Essay 16: Fish Anatomy 37
Essay 17: Fluorescent Tagged Proteins 38
Essay 18: Four Classes of Biological Macromolecules 39
Essay 19: Honeybees 42
Essay 20: LTP and Synaptic Transmission 43
Essay 21: Macronutrients, Dangerous Diets, and Starvation 45
Essay 22: Neuroscience Predictions 48
Essay 23: Rainforests of the World 49
Earth Science & Geology 52
Essay 24: Carbon-14 Dating 53

Essay 25: Climate Change over the Past Thousand Years 55
Essay 26: Geography of the Atlantic Ocean 58
Essay 27: Geological Processes in the Earth's Interior 60
Essay 28: Oceanic Heat Content 62
Essay 29: Rock Formation in Geology 63
Essay 30: Soil Quality – Coffee Grounds, Eggshells, and Earthworms 65
Human Anatomy & Physiology 67
Essay 31: Amino Acid Metabolism Pathways 68
Essay 32: The Amygdala 71
Essay 33 The Biological Basis of Memory 73
Essay 34: How does the brain sense osmolality and regulate water balance? 76
Essay 35: Cell Junctions 78
Essay 36: Cholesterol Explained 80
Essay 37: Diet and Metabolism 82
Essay 38: The Human Eye 84
Essay 39: Fat – So Much More Than an F Word ;-) 87
Essay 40: Glucose and Sodium Homeostasis 89
Essay 41: The Human Circulatory System 97
Essay 42: Overview of the Human Immune System 99
Essay 43: How do human bones age? 102
Essay 44: Mitochondria 104
Essay 45: Neurotransmitters 107
Essay 46: Pituitary Gland Function 109
Essay 47: Platelets 111
Essay 48: How are Proteins Converted to Amino Acids? It's all about Proteases 113
Essay 49: How Do Steroids Work in the Body? 115
Essay 50: Temperature Control Mechanisms of the Human Body 117
Essay 51: Vitamin K and the Coagulation Cascade 119

Essay 52: White Blood Cell Surface Receptors in Innate Immunity 122

Medicine & Disease 124

Essay 53: Alpha-1 Antitrypsin Deficiency 125

Essay 54: Alcoholism 127

Essay 55: Aging and the Brain Alzheimer's Disease 129

Essay 56: Common Causes of Anemia 131

Essay 57: Arthritis Explained 133

Essay 58: Autopsy Basics 135

Essay 59: Cardiomegaly 138

Essay 60: Chromosomal Aneuploidy 140

Essay 61: Colon Cancer Risk Factors 143

Essay 62: Common Causes of Anemia 145

Essay 63: Common Myths about Medical School Acceptance 147

Essay 64: Crohn's Disease 150

Essay 65: Diabetes Mellitus – Scrape the Icing Off the Cupcake 152

Essay 66: Diagnosis and Treatment of Emphysema 154

Essay 67: Different Strokes for Different Folks 156

Essay 68: Five Commonly Misdiagnosed Diseases 158

Essay 69: Foods that Help Prevent Heart Disease 161

Essay 70: Hashimoto's Thyroiditis 163

Essay 71: Heat Stroke 165

Essay 72: Huntington's Disease 167

Essay 73: Pathogenesis of Primary Hypertension 169

Essay 74: Hypertension – Secondary Causes 171

Essay 75: Hypertension: Non-Pharmacological Treatment Options 173

Essay 76: Leprosy 175

Essay 77: Treatment Options for Drug Resistant Leukemia 178

Essay 78: Lung Cancer – Diagnosis, Staging, and Treatment 180

Essay 79: Neuroblastoma 182

Essay 80: Neurofibromatosis / Von Recklinghausen Disease	184
Essay 81: Parkinson's Disease and the 1918 Flu Pandemic	186
Essay 82: Patau Syndrome	188
Essay 83: The Philadelphia Chromosome and Leukemia	189
Essay 84: Pituitary Tumors	191
Essay 85: Pneumonia - CAP vs. Nosocomial Infections	193
Essay 86: Portal Hypertension	195
Essay 87: Causes of Right Lower Quadrant Pain	197
Essay 88: Sickle Cell Anemia	199
Essay 89: Systemic Lupus Erythematosus	201
Essay 90: Sports Injuries	204
Essay 91: Stages of Chronic Kidney Disease	205
Essay 92: Staphylococcus – the Ultimate Superbug	207
Essay 93: Tumors of the CNS - Diagnosis and Treatment	209
Essay 94: Von Hippel-Lindau Disease	211
Pharmacology	**213**
Essay 95: Anti-Arrhythmic Drugs A Double-Edged Sword	214
Essay 96: Antibiotics	216
Essay 97: Caffeine	219
Essay 98: Ergot Alkaloids	221
Essay 99: Incredible Insulin	223
Essay 100: Melatonin	226
Essay 101: How Do Neurotoxins Work?	227
Essay 102: Courses to Take Prior to Pharmacy School	229
Essay 103: Quetiapine and the CDS Schedule	232
Essay 104: Dr. Katz's CVA Cocktail / X-Glut-5	234
Religion	**238**
Essay 105: Judaism - What is Apocryphal Biblical Literature?	239
Essay 106: Judaism - The Patriarchs	242

Essay 107: Judaism - The Star of David 244
Travel 246
Essay 108: A Whirlwind Tour of Destinations 247
Essay 109: Fall Foliage - MD, DC, VA 250
Essay 110: High School Memories: My Senior Class Trip To Israel 252
U.S. History & the American Experience 255
Essay 111: Economics 101: Dr. Katz Goes to Wall Street 256
From Commander in Chief to the Imperial Presidency 265
The Growth of Presidential Power from 1865-2014 265
Essay 112: Election 2016 273
Essay 113: U.S. Civil War - The Battle of Shiloh 276
World History & International Politics 278
Essay 114: The Dreyfus Affair 279
Essay 115: Geography and International Relations 281
Essay 116: German Military Blunders of WWI 288
Essay 117: From Hunter-Gatherers to Farmers 290
Essay 118: International Security in an Era of Globalization 292
Essay 119: The Legacy of the Thirty Years' War 299
Essay 120: The Seven Wonders of the Ancient World 303
Essay 121: Who Won the Cold War? 305
Essay 122: Was the League of Nations a Failure? 311
Essay 123: The Yuan Dynasty and Medieval China 315
Epilogue 317
References 318

Dedication

To my family, all of my friends, my pet Little Cat, my classmate Elina Segel, who passed away after a long battle with breast cancer; a recently deceased 98-year-old man I'll call Zaydee Z"L, and to my late, great cousin Alan Greenbaum a.k.a. Joey McFuzz, who once told me, "Life isn't over until you're flying with the angels in heaven." Fly free, Joey McFuzz. I miss you.

Preface The 4th Degree

This collection of essays represents my finest freelance writing for a website called Helium.com. Just for the record, the website was zapped into Cyber-oblivion at the end of 2014, probably because they paid their writers $3.00 per article on a good day. Luckily, I saved every major article I wrote for them. You're welcome, Helium.

On a deeper level, I wrote these essays because I LOVE to teach. Writing is the next best thing if I can't speak in front of a group of students or tutor in person. These essays represent the highlights of many long days of study (and countless caffeine-fueled nights) from my years of higher education at the Johns Hopkins University, Penn State University College of Medicine, the University of Alabama at Birmingham, and the University of Maryland School of Pharmacy.

You may wonder why I stopped after 4 degrees. Why not go to law school or get an MBA? Why not go to Yeshiva for a few years and become an ordained Rabbi?

Let's see…reading about legal cases literally puts me to sleep; all I need to know about business I learned on the job from some really successful business people plus watching Jim Cramer on Mad Money; and, last but not least, my parents refused to pay another cent for me to attend school after age 40. Dr. Katz had yet to stand on his own two feet.

As for becoming a rabbi, I hope Orthodox Jewish readers don't take this the wrong way, but I would go off the deep end if my wardrobe consisted exclusively of white shirts and black slacks. Also, I would need about 7 years to study the Talmud cover, – which I intend to do now that an online translation exists. Plus, eating for free at Bar Mitzvahs is tempting, so I'll keep the Rabbinic option on the table.

To all my readers: The joy of learning transcends any man-made degree. Never forget that!

So, let's begin.

Astronomy / Cosmology

Essay 1: Cosmology - A Brief History

Numerous volumes have been written about the science of cosmology over the centuries. Strange as it may sound, until the 1920s, many scientists thought our Milky Way galaxy constituted the entire universe. Terms now familiar to the general public, such as the Big Bang and expansion of the universe, did not even exist a single lifetime ago. This article will briefly discuss how modern cosmology grew out of astronomy, astrology, and physics, then touch on some key problems cosmologists are working on today.

One definition of cosmology is the quest to figure out the origin and overall structure of the universe. The roots of that quest are unknown but undoubtedly began in prehistoric times. Celestial bodies have always fascinated humans, notably the sun, moon, stars, and planets. To primitive people living millennia before the advent of science, it was no doubt easier to imagine celestial objects as gods or goddesses rather than mundane balls of rock or gas.

Nonetheless, by ancient times, various civilizations had specialists whose chief task was to observe the motions of the sun, moon, planets, and stars. Gradually, peoples as far apart in space and time as the Chinese, the Egyptians, the Greeks, and the Maya developed solar and lunar calendars. In ancient Egypt, the priests observed that the Nile flooded its banks every August, around when the Dog Star (Sirius) rose just before sunrise. Halfway around the world, the Maya developed two calendars – a solar calendar and another based on the orbital cycle of the planet Venus.

The Heliocentric View Triumphs: Copernicus and Galileo

From ancient times through the Middle Ages, most people, with the notable exception of Hipparchus of Rhodes, accepted a geocentric view of the cosmos (sometimes called the Ptolemaic model), which placed Earth in a privileged place at the center of the universe. By late medieval times, astronomy and astrology began to diverge in Europe. The first European astronomer to break with the geocentric view was the Polish priest Nicolaus Copernicus (1473-1543), whose work, On the Revolutions of the Celestial Orbs, was published posthumously. Most scholars agree that its heliocentric view inspired the work of an even more famous scientist, Galileo Galilei (1564-1642).

Galileo had one advantage that no person before him enjoyed: the telescope, first constructed in the year 1609. Although his telescope was crude by today's standards, it allowed Galileo to see mountains on the moon, four small satellites orbiting the planet Jupiter, the rings of Saturn, and countless tiny stars that form the band of light called the Milky Way. Although the Catholic Church put Galileo under house arrest for publishing his observations, there was no way to reverse the invention of the telescope. By the late 17th century, astronomers throughout Europe were expanding upon the work of Galileo as well as that of the German astronomer Johannes Kepler, who formulated the basic equations describing planetary orbits.

The Enlightenment: Isaac Newton

It is difficult to exaggerate the importance of Isaac Newton (1642-1727) in the development of mathematics, physics, and cosmology. Newton's theory of gravitation made the laws of physics applicable to the realm of astronomy. Newton's work in the field of optics heralded the development of more powerful telescopes. Finally, Newton's discovery of calculus shaped the work of all the great scientists who followed him, including Albert Einstein (1879-1955).

The Early 20th Century: Einstein's Theory of Relativity

In 1905, Einstein published his special theory of relativity, whose core theme is that the speed of light constitutes the highest speed attainable in our universe. By 1916, Einstein published his general theory of relativity, in which he reformulated Newton's theory of gravity. Space and time are separate entities in the Newtonian universe, independent of all other natural forces. Einstein's revolutionary theory transformed this model into a universe embedded in a four-dimensional space-time matrix. Massive bodies like stars and planets do not exert gravity as some mysterious force acting at a distance; rather, they warp the fabric of space-time itself, and this effect is felt as gravity. Time is not an absolute entity but a dimension inseparable from space. As profound as these concepts are, the theory raised an even more amazing prediction – the universe cannot exist in a static state; rather, it must be expanding or contracting. Little did Einstein know a few decades later, an astronomer named Edwin Hubble would prove this idea was fundamentally correct.

1920 Curtis-Shapley debate:

Heber Curtis believed the solar system to be at the center of the Milky Way galaxy but that many so called nebulae like Andromeda were actually galaxies in their own right. Harlow Shapley argued that given the distribution of stars visible from Earth with the brightest region of the Milky Way in the constellation Sagittarius, the solar system must be located in an outer arm of the Milky Way galaxy. However, he was convinced that the Milky Way encompassed the entire visible universe (considering that in 1920 the accurate distance between our galaxy and nebulae like Andromeda remained a matter of debate).

Curtis was incorrect about the location of our solar system within the Milky Way but correct about Andromeda and many nebulae existing as galaxies far beyond our own. Shapley was incorrect that the Milky Way contained the entire visible universe but was correct in establishing the location of our solar system in a spiral arm rather than at the galactic center. As it turns out, the solar system is slightly closer to the Milky Way's center (26,000 light years away) compared to the galaxy's edge (using an approximate diameter of 105,000 light years).

Edwin Hubble – the Expanding Universe and the Big Bang

Today, the name Hubble brings to mind the orbiting space telescope launched in 1990. The astronomer for whom it was named, Edwin Hubble (1889-1953), first came to prominence in the astronomical community when he established that the Andromeda nebula was actually a galaxy in its own right, not merely a patchy cloud within our Milky Way galaxy. Soon after, Hubble and his contemporaries observed that many objects once thought to be nebulae were actually galaxies quite distant from our own.

Several years later, Hubble noticed an astounding pattern: almost all observable galaxies displayed a red shift in their spectra, meaning that light from these galaxies was being stretched, or shifted, toward the red end of the spectrum. Much as the sound from a siren becomes stretched out as an ambulance rushes away from an observer, the light from distant galaxies is also stretched to lower frequencies of the spectrum (the red end) because they are moving away from our own galaxy. On top of that, Hubble noted that the more distant the galaxy, the larger its red shift, and the faster it receded from the Milky Way.

The conclusion was inescapable (although it took some scientists decades to accept): at some point in the distant past, all matter in the universe was compressed into a single point – an incredibly dense, unimaginably hot fireball. Around 15 billion years ago, our universe burst into existence from this primordial fireball, an event now popularly termed the Big Bang.

Modern Cosmology – the Search for the Hubble Constant and the Fate of the Universe

Since the 1960s, many cosmologists have tried to measure the universe's expansion rate in an attempt to learn its age and the ultimate fate of the cosmos.

The term for the universe's expansion rate, the Hubble Constant, is denoted H_o. The value of the Hubble Constant is estimated at 72 kilometers per second per megaparsec. This means that with each passing second, 72 km of space opens up over the incredibly vast distance of 3 million light years. Keep in mind a single light year is equal to 5.9 trillion miles. Because of gravitational attraction, the expansion of the universe is undetectable within our own galaxy (100,000 light-years across) and difficult to discern even within clusters of galaxies. Only on the tremendously vast distance scale of galaxies more than 100 million light years away from us does the expansion of the universe become noticeable.

Knowing the value of Ho and whether it has changed over time is the key to predicting whether the universe will expand forever and eventually burn out or stop expanding and ultimately collapse into another fireball. Today, many astronomers are convinced the expansion of the universe is accelerating for unknown reasons. This has led them to concoct ideas such as dark energy – a sort of antigravity – to account for this bizarre phenomenon. Other cosmologists remain unconvinced.

Their theories invoke the concept of a 'multiverse' – the idea that our universe is one of an infinite number of universes that separate and coalesce over unimaginably long periods. As with most profound questions, the debate will likely continue for the foreseeable future.

Essay 2: Constellations Cassiopeia

The constellation Cassiopeia, the Queen, is found in the northern sky, relatively close to the north celestial pole. This constellation is circumpolar for all sky watchers who live above 40 degrees north latitude. Cassiopeia borders the constellations Cepheus (the King), Camelopardalis (the Giraffe), Perseus (the Hero), and Andromeda (the Princess).

In Greek mythology, Cassiopeia was the queen of Ethiopia. When she boasted that her daughter Andromeda was more beautiful than the sea nymphs, Poseidon sent a sea monster (identified with the constellation Cetus) to ravage the coast. Out of desperation to save her country, Cassiopeia allowed her daughter Andromeda to be chained to a rock as a sacrifice to the sea monster, but at the last moment, Andromeda was rescued by the hero Perseus. As a punishment for her arrogance, Cassiopeia was placed in the sky on a throne that faces downward for half of its celestial course.

When and Where to Look

Cassiopeia is visible on any clear night of the year from most locations in the northern hemisphere. For amateur stargazers, September through March are the optimal months for spotting Cassiopeia. In the southern hemisphere, the constellation is visible from May to August at locations north of the Tropic of Capricorn (23.5 degrees south latitude).

The easiest way to identify this constellation is to find the North Star (Polaris) at the end of the Little Dipper's tail. The Little Dipper is the most noticeable feature of the constellation Ursa Minor. As a reminder, Polaris is always as many degrees above the northern horizon as you are north of the equator. Look approximately two-hand widths or so away at the area of the sky opposite the Little Dipper, and Cassiopeia should be visible as an M, W, E, or 3 shape depending on the season of the year and/or the time of night you are star gazing.

Stars

Cassiopeia contains no first-magnitude stars. It does contain several 2nd and 3rd-magnitude stars, and its shape is rather distinctive. Schedar, meaning breast in Arabic, is the alpha star in the constellation. It is magnitude 2.2 and 120 light years from Earth. The beta star in Cassiopeia is called Caph, meaning hand in Arabic and Hebrew.

It is magnitude 2.3 and located 42 light years from Earth. The gamma star is called Cih, meaning whip in Chinese. It is 780 light years away, and its brightness varies sporadically from magnitude 1.6 to 3.0.

Deep sky objects

Cassiopeia contains two Messier objects, M52 and M103. Both are 7th magnitude open star clusters, visible through a small telescope or binoculars on a clear night. The planetary nebula NGC 7635, sometimes called the Bubble Nebula, is fainter and best viewed through a medium size telescope The elliptical galaxy NGC 185 is faint at 10th magnitude and requires a medium size telescope to be seen. The Milky Way runs directly through Cassiopeia and is easily visible through binoculars or a small telescope.

Essay 3: Identifying Constellations for Amateur Stargazers

One of the challenges amateur star observers face is learning to identify constellations. When the average person hears that the sky is divided into 88 constellations, his/her initial reaction is often, "Forget it. It's hopeless. I'll never remember most of them." The purpose of this article is to convince aspiring star gazers otherwise. Identifying a handful of constellations is the first step (and arguably the most important one) in learning to navigate around the night sky.

The Big Dipper

For observers in the Northern Hemisphere, a good place to start is finding the Big Dipper, the most noticeable group of stars in the constellation Ursa Major (the Great Bear). Under clear skies, the Big Dipper is visible at some time of night from all locations north of the equator. One way to identify the Big Dipper is to find north, using a compass if necessary. Look for seven stars that form the pattern shown on the diagram. You will notice that the Dipper's shape somewhat resembles a ladle or large spoon, and most of its stars are second and third magnitude, meaning they are fairly bright.

To an observer facing north, the Dipper appears to move in a counterclockwise circle around the north celestial pole. The north pole of the sky is marked by the second magnitude star Polaris, better known as the North Star. Once you are comfortable finding the Big Dipper, you can use it as a jump off point to find many other constellations as the seasons pass.

The Spring Sky: Arc to Arcturus, and speed on to Spica, but don't miss Leo.

The Big Dipper is directly over Polaris for most of spring. If you follow the curve of the Dipper's handle, it will point to a bright orange star called Arcturus in the constellation Bootes, the Bear Driver. Extending the same line past Arcturus will lead you to the bright bluish white star Spica in the constellation Virgo.

Northwest of Virgo is the constellation Leo, the Lion. Leo can also be found by drawing a line backwards through the pointer stars of the Dipper's bowl. The line will end near Regulus, a bluish white first magnitude star that marks the Lion's heart.

The Summer Sky: The Summer Triangle, the Scorpion, and Sagittarius

One highlight of the summer sky is three constellations whose brightest stars form a large upside down triangle in the sky. These stars are Vega in the constellation Lyra (the Harp), Deneb in the constellation Cygnus (the Swan), and Altair in the constellation Aquila (the Eagle). Southwest of Altair (the farthest south of the three stars of the summer triangle) is the constellation Scorpio with the bright red star Antares at its center. Scorpio does resemble a scorpion and is visible in the southern sky all summer long. East of Scorpio is the constellation Sagittarius, the Archer. This constellation contains dozens of deep sky objects visible through binoculars or a small telescope. The center of the Milky Way galaxy is located in Sagittarius.

The Autumn Sky: The Great Square of Pegasus, Auriga, and Taurus

The Great Square of Pegasus rises in the northeast during late summer. Due north of Pegasus is a hazy patch of light. This is M31, the Andromeda Galaxy, one of the few galaxies visible to the naked eye. Although it is 2.5 million light years away, the Andromeda Galaxy and milky Way are both part of the same galactic cluster known as the Local Group. The next noticeable constellation to rise after Pegasus is Auriga, the Charioteer, with its bright yellow star Capella. An hour or two after Auriga rises, Taurus the Bull becomes visible in the eastern sky. This constellation contains the bright red star Aldebaran, and two naked eye star clusters, the Hyades, which form the V shape of the Bull's horns, and M45, a.k.a. the Pleiades (the Seven Sisters).

The Winter Sky: Gemini, Orion, and the Dog stars

The rising of Taurus tells us that the next constellation in the zodiac will rise soon. This is Gemini, the Twins. The two brightest stars in Gemini are named Castor and Pollux. Gemini is visible from autumn through the following June for observers in the Northern Hemisphere. The main highlight of the winter sky, however, is Orion (the Hunter), located south of Taurus and Gemini.

Orion straddles the celestial equator and can be seen from most locations on Earth for at least 6 months out of the year. Orion contains two first magnitude stars, the red variable star Betelgeuse, which means House of the Twins in Arabic, and the blue giant star Rigel, which means foot in Arabic and Hebrew. Orion's Belt contains three fairly bright stars, occasionally called the three kings, which divide the constellation neatly in half. Below the second star in the Belt is a group of stars called Orion's Sword. The middle star of the Sword appears fuzzy to the naked eye. That's because it is really a cluster of stars called M42, the Great Orion Nebula.

A line drawn through Orion's belt to the southeast points to the constellation Canis Major (the Large Dog). Canis Major contains Sirius (the Dog Star), the brightest star in the night sky at magnitude -1.47. Sirius is a brilliant white star visible from all latitudes south of Greenland.

North of Sirius, just above the celestial equator, is Canis Minor (the Small Dog), which contains the bright yellow star Procyon, which literally means "before the dog", as it rises a short time before Sirius.

Southern Hemisphere

Key constellations for observers in the Southern Hemisphere to learn include Carina (the Ship's Keel), which contains Canopus, the second brightest star in the night sky; Centaurus (the Centaur), which contains the first magnitude stars Alpha and Beta Centauri; and the nearby constellation Crux, the Southern Cross, which contains two first magnitude stars, Acrux and Becrux. One other first magnitude star visible from latitudes south of Florida is Achernar, which marks the southern end of the constellation Eridanus (the River), southwest of Orion.

The Large and Small Magellanic Clouds, named after the 16th century explorer Ferdinand Magellan, are satellite galaxies of the Milky Way. They are located in the faint constellations Dorado (the Goldfish) and Tucana (the Toucan), respectively. The Large Magellanic Cloud contains NGC 2070. Better known as the True Lovers' Knot or the Tarantula Nebula, it is one of the brightest star clouds in the known universe. If this nebula were located 1,000 light years from Earth (as opposed to 70,000 light years away), it would be as bright as the full moon and cover 60 times more area in the night sky.

Essay 4: How to Use Constellations to Tell Time

Before the advent of watches and clocks, groups of people throughout the world used the rising and setting of stars to mark off the hours of the night. Although timekeeping devices are ubiquitous in the modern world, telling time by the stars is still an enjoyable activity and, in certain situations, a valuable skill. This essay will focus on easy-to-find stars and constellations, with an emphasis on patterns visible to observers in the Northern Hemisphere.

One of the most easily recognizable constellations for beginners is Ursa Major, the Great Bear. The most noticeable part of the Great Bear is an asterism, or pattern, known as the Big Dipper. Most of the stars in the Big Dipper are of second and third magnitude, meaning they can be seen even under hazy conditions or in areas with substantial light pollution. So that the next few paragraphs make sense, please take a moment to refer to the diagram of the Big Dipper in relation to Polaris, the North Star, which marks the tail of Ursa Minor, or Little Dipper. Finding north is not complicated, assuming you have a compass or take notice of the position of sunrise or sunset.

In the early evening in spring, the Big Dipper is directly above the North Star, just as it appears in the diagram. Six hours later, the Big Dipper has moved 90 degrees and is now to the left, or due west of the North Star. Six hours after that (right around dawn), the Big Dipper is directly below the North Star. In twelve hours, the Big Dipper has moved through half of a great circle with the North Star in the center.

In the winter, the Big Dipper lays to the right, or due east, of the North Star. Twelve hours later, it has reached its summer position. In the span of one hour, stars move 15 degrees to the west in a counterclockwise direction relative to the North Star. 15 degrees is approximately the distance from your thumb to the tip of your pinkie with your arm outstretched.

For more advanced sky watchers, as long as you know the season of the year, you can usually estimate the time of night accurately from the positions of a few bright stars. In late winter and early spring, Arcturus in Bootes can be found in the eastern sky by following the curve of the Big Dipper's handle. Arcturus is a bright orange star and is relatively easy to spot.

A bright blue star called Vega rises in the northeast four hours later. Alternatively, you can look to the northwest and spot the Pleiades and Taurus setting in the early evenings of March and April.

In summer, three stars called the Summer Triangle dominate the night sky. They are Vega, in Lyra, the Harp; Deneb, in Cygnus, the Swan; and Altair, in Aquila, the Eagle.

Other noticeable constellations in the summer sky form part of the zodiac: Leo, the Lion in the southwest; followed by Virgo, the Virgin; then the dim constellation Libra, followed by Scorpio, the Scorpion, with its bright red star Antares. Sagittarius, the Archer, rises in the southeast a few hours after the Scorpion.

The coming of autumn is marked by two bright stars: Capella, in the constellation Auriga, the Charioteer, and Fomalhaut, the only bright star in Piscis Austrinus, the Southern Fish. The important thing to remember is that although these two stars are located in the extreme northeast and southeastern parts of the sky, they rise roughly at the same time. Fomalhaut set around 2 a.m., around 2 hours before Capella has reached the halfway point across the sky. Another constellation easily recognized in the northeastern sky in late summer and early fall is Pegasus, the Winged Horse, marked by four stars called the Great Square. Pegasus is visible throughout the winter months, setting by early spring.

The constellations Orion, the Hunter, and Canis Major, the Big Dog, dominate the winter sky. Orion contains the bright red star Betelgeuse and the brilliant blue star Rigel. Three stars marking the hunter's belt run midway between these two along the sky's celestial equator. Southeast of Orion is Canis Major, which contains Sirius, the brightest star in the night sky. By December, Orion is visible just after sunset and sets around sunrise. Sirius is visible throughout winter to Northern Hemisphere observers and sets by early May.

Southern Hemisphere

For observers south of the equator, many constellations will still be visible, with the notable absence of the Big and Little Dippers. Constellations straddling the celestial equator, like Orion, are useful guides for at least 6 months or so each year. Certain constellations are circumpolar in relation to the south celestial pole, meaning they can be seen on any clear night of the year. The ones with bright stars include Centaurus (the Centaur), which contains Alpha and Beta Centauri; Crux, the Southern Cross, with the stars Acrux and Becrux; Carina, the Keel, which contains Canopus, the second brightest star in the night sky; and the southern region of Eridanus, the River, with its bright star Achernar.

In terms of position, Achernar is southwest of Orion. Canopus is almost directly due south of Sirius, the Dog star. The Southern Cross and the Centaur rise around 6 hours after Orion, so Orion is close to setting by the time these constellations become visible. As a final tip, remember that the Scorpion rises an hour or so after Orion sets. Conversely, an hour or so after the Scorpion sets, Taurus the Bull, with its bright star Aldebaran, rises.

Essay 5: The Constellation Cygnus

The constellation Cygnus, the Swan, is located in the northern sky between the constellations Lyra (the Harp), Draco (the Dragon), Cepheus (the King), Lacerta (the Lizard), and Vulpecula (the Little Fox). Cygnus is one of the few constellations that bears some resemblance to its namesake. Its brightest stars form an asterism, or pattern, better known as the Northern Cross. In Greek mythology, Cygnus corresponds to the swan in the story of Zeus and Leda, in which Zeus disguised himself as a swan, seduced the queen, and became the father of Pollux and Helen of Troy. This article will discuss the easiest ways for sky watchers to find Cygnus and then explore some of the stars and deep sky objects located in this constellation.

When and Where to Look

Cygnus is visible at some time of night nearly all year long to observers in the Northern Hemisphere. Cygnus rises in the northeast after midnight during spring. It can be seen overhead during the summer months in the early evening and in the northwestern sky throughout autumn. By late December, Cygnus sets a couple of hours after sunset. For most of the winter, Cygnus can be seen shortly before sunrise and then after midnight by early spring.

Stars

Two of the most noticeable stars in Cygnus are Deneb and Albireo. Deneb, which means tail in Arabic, is a blue supergiant star that marks the tail of the swan or upright of the Northern Cross. Deneb is the only first-magnitude star in Cygnus and 19th among the 20 brightest stars visible in the night sky. In terms of absolute brightness, however, Deneb is one of the brightest stars in our galaxy, over 60,000 times brighter than our sun. Deneb is so massive that it radiates as much light in a single day as our sun does in a century.

The reason Deneb does not appear brighter is due to its immense distance from Earth, estimated at 1,600 light years. A light year is the distance light travels in one year, or approximately 5.8 trillion miles. In other words, Deneb is so far away that the light reaching your eyes today left the star not long after the Visigoths sacked Rome in the year 410 AD.

Albireo, meaning the hen's beak, marks the head of the swan or base of the Northern Cross. To the naked eye, Albireo is a third-magnitude star, somewhat dimmer than the three stars marking the Swan's body. A small telescope reveals Albireo to be a blue and gold double star. Astronomers estimate these two stars to be 380 light years from Earth. Albireo is a physical double, meaning its component stars are locked into orbit around each other by mutual gravitational attraction.

Several deep sky objects in Cygnus are visible through a telescope. They include the Veil Nebula, a supernova remnant; three emission nebulae, the Crescent Nebula, North America Nebula, and Pelican Nebula; as well as the Messier objects M29 and M39, both of which are open star clusters. The Milky Way runs directly through Cygnus and is easily visible through binoculars or a small telescope.

Essay 6: Edwin Hubble and the Hubble Constant

Most people have heard of the Hubble Space Telescope (HST), but relatively few people are familiar with the astronomer after whom it was named. Edwin Hubble (1889-1954) was a key figure in 20th-century astronomy, so much so that not only was a space telescope named for him but a constant of physics as well. This article will explore the highlights of Hubble's work and its significance for modern-day astronomers and cosmologists.

Hubble began his career as an astronomer in the early 20th century. At the time, three questions dominated the field without consensus. First, is our solar system located at the center of the Milky Way galaxy or somewhere in the periphery? Second, what is the size of the Milky Way in light years? Finally, arguably most important, is our galaxy the only one in the universe, or do other galaxies exist outside our own?

In 1920, two leading astronomers, Shapley and Curtis, debated these questions at a famous forum held in Washington, DC. Predictably, the debate generated more heat than light. Unbeknownst to all concerned, Hubble would tackle these questions over the next two decades. In the process, he forged a permanent link between the sciences of astronomy and cosmology.

Hubble's first breakthrough was the detection of stars called Cepheid variables in the Andromeda 'Nebula.' These stars fluctuate in brightness in a predictable cycle of months to years. Compared to their counterparts in the Milky Way, the Cepheids in Andromeda cycled far more slowly. This confirmed that Andromeda was not a gas cloud within our own galaxy but rather a spiral galaxy located far beyond the Milky Way – some 2.5 million light years away. As larger telescopes became available during the 1920s, astronomers discovered hundreds then thousands of galaxies separate from the Milky Way.

Discovering that our galaxy is one of a multitude of galaxies was a tremendous feat in and of itself. Hubble's next discovery was truly monumental. By 1929, he noticed that the emission lines of nearly all observable galaxies were shifted or stretched toward the red end of the electromagnetic spectrum. This meant that, aside from gravitationally bound clusters of galaxies, most galaxies are moving away from one another. Then Hubble discovered a trend that changed our view of the universe forever.

The more distant a galaxy was from the Milky Way, the larger its red shift and the faster it was receding. Essentially, galactic red shifts were the first solid evidence of an expanding universe.

While astronomers (including Hubble) struggled to make sense of these observations, it was not long before cosmologists took Hubble's discovery to its logical conclusion: if galaxies throughout the universe are rushing away from each other, at some point in the distant past, they must have shared a common origin – in an unimaginably hot, dense ball of matter, energy, space, and time that scientists call the Big Bang.

So, how long ago did the expansion of the universe begin? In other words, when was the universe born? That all depends upon how fast the universe has expanded from the Big Bang up to the present, captured by the deceptively simple equation Velocity = Distance X Rate constant. Appropriately enough, the rate term was named the Hubble Constant, denoted Ho.

The Hubble Constant is technically not a constant but rather a rough estimate of the universe's expansion rate. The units of Ho are kilometers per second per megaparsec. One parsec equals just over 3 light-years; a megaparsec equals over 3 million lightyears. Most astronomers think the value of Ho falls between 50 and 100 km/s/Mpc. This means that with each passing second, between 50 and 100 kilometers of intergalactic space opens up over the vast distance of 3 million light years.

The reciprocal of Ho gives the approximate time elapsed since the expansion of the universe began. If the current estimate of Ho (68 km/sec/Mpc) stands correct, this corresponds to a universe between 12 and 15 billion years old. According to astronomers, the expansion of the universe is most noticeable in the gigantic voids between galaxies and galactic clusters. Within individual galaxies such as the Milky Way, the gravitational attraction between stars and other matter counteracts the force of expansion; otherwise, galaxies would be ripped apart or would never have formed in the first place.

Interestingly, astronomers cannot account for the majority of matter responsible for gravitational attraction within the Milky Way or other galaxies. This led them to propose the existence of 'dark matter,' a heretofore uncharacterized substance that comprises most of the matter in the universe. Some astronomers think that particles called neutrinos account for the majority of dark matter. Others invoke the existence of weakly interacting massive particles or so-called WIMPS. Still, others think the presence of supermassive black holes at the center of most galaxies may account for dark matter and its gravitational effects. The debate continues.

Ho is technically not a 'constant' because, in the very early universe, the expansion rate may have been higher before the formation of the first stars and galaxies. Once the universe evolved from a hot, high-energy state into a cooler one where matter could aggregate, gravity took hold over large distances, and the universe's expansion rate slowed down.

In recent years, however, some astronomers think they have discovered a strange trend – the expansion of the universe may actually be accelerating. They attribute this acceleration to the existence of 'dark energy,' the mysterious counterpart of so-called dark matter. Others are not convinced, pointing out that if a presumably tangible entity like dark matter remains so elusive, how can we ever expect to confirm the existence of dark energy? As with all debates in modern astronomy, this one is likely to continue for quite some time.

Essay 7: How Space and Time Became Space-Time

Today, the term 'space-time matrix' is well known to scientists and laypeople alike. Until the early 20th century, however, almost everyone held the classical view that space and time exist as two absolute, independent entities. In 1915, Einstein's general theory of relativity transformed that view forever. This article will explore how this radical transformation occurred.

For most of human history, almost no one conceived space and time as mutable or interrelated. This is no accident. In our everyday lives, all events seem to occur in the context of rigid three-dimensional space and the one-way arrow of time. Isaac Newton captured this view in his epic work The Principia when he wrote, "Absolute, true, and mathematical time, in and of itself and of its own nature, without reference to anything external, flows uniformly and by another name is called duration."

Newton's sentiments aside, there does seem to be a profound connection between space and time. If your friend asks you to meet at 8 pm, the next obvious question would be, where? Conversely, if your friend asks you to meet at a local restaurant, you would instinctively ask, at what time? Beyond this obvious link, however, nothing in our everyday experience seems capable of altering space or time itself. The notions that space can curve like fabric or that a strong gravitational field can alter the flow of time seem bizarre and antithetical to common sense.

The first indication that the Newtonian theory of gravity was incomplete came from observations of the orbit of the planet Mercury. Astronomers noticed that the planet's perihelion point, or location where Mercury is closest to the sun, gradually shifts over long spans of time. Nothing in Newtonian physics accounts for this anomaly. While mathematicians like Gauss had developed non-Euclidian systems of geometry by the early 19th century, concepts like spherical curvature seemed to apply to tangible bodies like the Earth and moon. No one seriously entertained the possibility that space itself might be curved.

When Albert Einstein published his Theory of Special Relativity in 1905, he purposely shied away from addressing gravity. Instead, the theory focused on the speed of light as the maximum possible speed in the universe and its implications for the laws of physics. A decade later, Einstein published his Theory of General Relativity. One of its main themes is that gravity does not act as an invisible string but rather as a force field akin to electromagnetism. It directly contradicted the Newtonian view of gravity as "force acting at a distance." This idea was vindicated by the solar eclipse of May 1919. Arthur Stanley Eddington, a proponent of Einstein's theory, traveled to islands off the coast of West Africa. During the eclipse, he took photographs showing a shift in the predicted position of stars that became visible for a few minutes during totality, when the moon completely covered the sun's disk.

It took many more years and experiments to demonstrate the time dilation effect predicted by General Relativity. Essentially, this idea states that the curvature of space affects the flow of time.

In the absence of a strong gravitational field, time slows down (or dilates) from the perspective of a non-moving observer. In a thought experiment, two twins are separated at birth. One twin is placed on a rocket ship while the other is left on Earth. The rocket travels at close to light speed and returns to Earth a year later. Upon arrival back on Earth, the baby on the rocket ship would be one year old, while his or her twin would be nearly 100 years old. Although this sounds like science fiction, records from atomic clocks placed on moving aircraft compared to identical instruments left on the ground demonstrate that time dilation is indeed real.

In spite of the vast progress we've made in understanding space and time, much more remains to be learned. When asked to define space and time, Einstein allegedly answered, "Space is what you measure with a ruler; time is what you measure with a clock." To paraphrase a professor of mine, at the end of the day, two mysteries, space and time, have been condensed into a single mystery - space-time.

Essay 8: The Constellation Leo

Leo, the Lion, is one of the twelve constellations of the zodiac. It is one of the few constellations that actually resemble the object or animal for which it is named. In Greek mythology, Leo was identified as the Nemean lion, a beast no weapon could harm, which Hercules strangled as the first of his twelve labors. Leo is located between the constellations Cancer and Virgo but can be found most easily by extending an imaginary line backward through the pointer stars of the Big Dipper's bowl.

Leo is visible before midnight to northern hemisphere sky observers from late autumn until the following July. This article will discuss some of the highlights of this constellation, including its brightest stars as well as deep sky objects visible through a telescope.

It is easiest for amateur skywatchers to find Leo in relation to the Big Dipper. In the late fall and winter months, Leo rises in the eastern sky to the right of the Dipper. During spring, Leo is overhead, and by early summer, Leo is setting in the west to the left of the Dipper.

Once you have located Leo, you will notice that its brightest stars form two distinct shapes: an isosceles triangle that marks the Lion's hindquarters and a backward question mark, sometimes called the Sickle, which forms the Lion's head. The bluish-white star at the end of the Sickle is Regulus (alpha Leonis). The name Regulus means prince, and centuries ago, astrologers designated it one of the four "royal stars" along with Aldebaran, Antares, and Fomalhaut.

At magnitude 1.36, Regulus is the brightest star in Leo and the twenty-first brightest star in the night sky. Regulus is located 77 light years from Earth and, as a blue giant star, shines 140 times brighter than our sun. In terms of apparent magnitude, the next brightest stars in Leo include Denebola (beta Leonis), which marks the Lion's tail; Algieba (gamma Leonis), in the Lion's mane; and Zosma (delta Leonis), located northeast of Denebola at the apex of the triangle.

Deep Sky Objects

Leo contains several deep sky objects, including M65, M66, M95, M96, and M105. The 'M' prefix was chosen in honor of Charles Messier, an 18th-century French astronomer who compiled a list of star clusters, nebulae, and galaxies that amateur astronomers often confused with comets. All of the Messier objects in Leo are spiral galaxies except for M105, which is an elliptical galaxy. None of these galaxies are visible to the naked eye but can be seen with a medium or large telescope.

Essay 9: Messier Marathon

The 110 Messier objects were named in honor of Charles Messier (1730-1817), an 18th-century French astronomer who devoted most of his career to discovering comets. Messier and his students spent a great deal of time distinguishing comets from other deep-sky objects. At the time, telescopes were still rather primitive, and photographic film did not exist. Over several years, Messier cataloged a collection of deep sky objects that amateur astronomers might easily mistake for comets. With certain obvious exceptions like M45 (the Pleiades), most Messier objects are not visible to the naked eye and may be confused for the halo of a comet when seen through a telescope.

Since Messier lived in France, all the objects in his catalog are visible from mid-Northern latitudes at some time of the year. During a 10-day window in late March, all 110 M-objects can be seen in one night if the weather cooperates and you have the stamina to observe the sky for 12 hours straight. Astronomy clubs coined the name Messier Marathon to highlight this annual event. When I lived in Birmingham, Alabama, I promised myself I would attend a Messier marathon. I suppose I'll have to add this event to my bucket list.

The Messier catalog can be divided up in several ways. For example, nearly all Messier objects consist of nebulae, star clusters, or galaxies. This makes sense, given that 18th-century telescopes were primitive by today's standards. I find the most useful method to be grouping Messier objects by apparent brightness: those visible to the naked eye, objects visible in binoculars, and finally, those objects best viewed through a telescope.

A few Messier objects can be seen with the naked eye, especially under clear, dark skies. These include M5, a globular cluster in Serpens; M31, the Great Andromeda Galaxy; M42/M43, the Orion Nebula; and M45, the Pleiades. The Andromeda Galaxy is the most distant object visible to the naked eye at some 2 million light years away. Objects visible through binoculars include M13, a globular cluster in Hercules; M44, known as Praesepe or the Beehive cluster in Cancer; and several clusters in Sagittarius.

Certain Messier objects, such as the Orion Nebula, appear magnificent in a telescope. At the nebula's center is the Trapezium, which appears as five tiny stars surrounded by a bright blue wispy cloud. This stellar nursery harbors some of the youngest stars in the Milky Way galaxy. Other objects, such as M1, the Crab Nebula in Taurus, or M57, the Ring Nebula in Lyra, seem far less impressive - until you recall that hours of photographic exposures must be taken to produce the colorful images seen on the chart below.

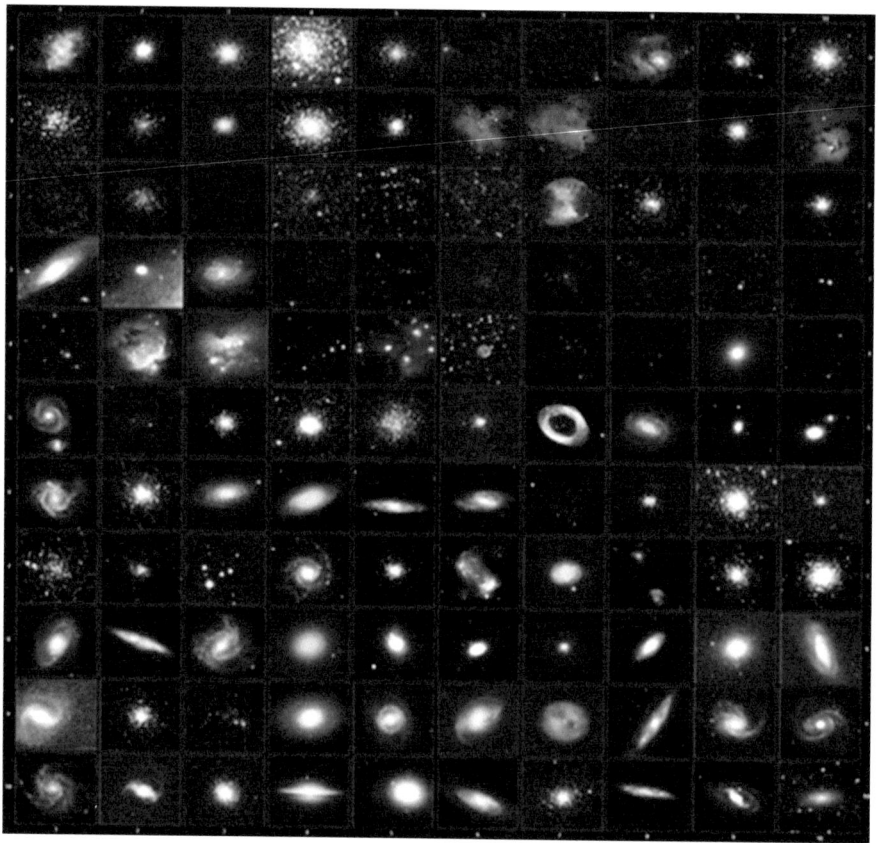

Essay 10: Orion the Hunter

Orion, the Hunter, is perhaps the most widely recognized of all constellations. In Greek mythology, Orion was a mighty hunter who triumphed over the mightiest beasts but fell victim to the bite of a lowly scorpion. To acknowledge this, the gods placed Orion on the opposite side of the sky as the constellation Scorpio so the two would never be visible simultaneously. To the ancient Egyptians, Orion was the god Osiris, while Amazonian Indians saw Orion as a giant river turtle. This article will explore some of the stars and deep-sky objects found in this fascinating constellation.

Orion straddles the celestial equator and is visible to skywatchers throughout the world for over six months out of each year. In mid-northern latitudes, Orion rises before midnight, starting in late October, and can be seen until the following April. The constellations bordering Orion include Taurus (the Bull) to the northeast, Gemini (the Twins) to the northwest, Canis Minor and Canis Major (Orion's Dogs) to the east and southeast, Lepus (the Hare) immediately to the south, and Eridanus (the River) to the southwest.

Stars

The brightest star in Orion is the blue supergiant Rigel, but for some obscure reason, the red star Betelgeuse was given the alpha designation. Rigel (which means 'foot' in Arabic and Hebrew) is located close to 800 light years away from Earth and marks the Hunter's left leg. Rigel appears bluish-white because its surface temperature is extremely hot – some 11,000 degrees Kelvin compared to 5,800 K for our sun. Rigel is also 17 times more massive than our sun and radiates over 60,000 times as much energy. Blue stars like Rigel have relatively short life spans as stars go – they exhaust their fuel in around 10 million years compared to 10 billion years for yellow stars like our sun.

Betelgeuse (Arabic for 'house of the twins') is a red supergiant estimated to be 640 light years away. Betelgeuse marks Orion's right arm or shoulder. The light output from Betelgeuse varies somewhat unpredictably over the course of several years, although Betelgeuse has always remained a 1st magnitude star.

This variability in brightness is common among red giants and supergiant's, dying stars that have exhausted most of their core hydrogen and now appear red because their bloated outer layers are cooler than those of blue, white, or yellow stars. If Betelgeuse were located in the same position as our sun, its surface would extend past the orbit of Mars.

The gamma star in Orion is named Bellatrix, a queen of the Amazons in Greek mythology. Bellatrix is a second-magnitude star estimated to be 240 light years from Earth. Bisecting the constellation of Orion is the Belt, made up of three second-magnitude stars named Mintaka, Alnilam, and Alnitak. These stars are occasionally called the three kings. Mintaka lies almost exactly on the celestial equator. Directly south of Orion's Belt is a group of three stars referred to as Orion's Sword, discussed in the next section.

Deep Sky Objects

Orion contains several deep sky objects, the most famous of which are M42 and nearby M43. Together, M42 and M43 comprise the Great Orion Nebula, located approximately 1,400 light years from Earth. To the naked eye, the Orion Nebula appears as a fuzzy star in the middle of Orion's Sword. A pair of binoculars reveals a blue cloud containing several tiny stars. These stars are sometimes called the Trapezium and are believed to be only a few million years old - extremely young in stellar terms. In a telescope, the Orion Nebula is a truly magnificent site. The other Messier object in Orion is M78, a diffuse reflection nebula located about 1,600 light years away.

To the southeast of Orion's belt is another famous deep sky object, the Horsehead Nebula (official designation: International Catalog 434). Unlike the bright Orion Nebula, however, the Horsehead Nebula is a dark dust cloud, which causes it to stand out in stark contrast against its brighter surroundings. The stunning images of the Horsehead Nebula are obtained by long exposure photography; a telescopic view of this object appears far less impressive.

Essay 11: The Sun, Moon, and Tides

For centuries, people have drawn a connection between the orbit of the moon and ocean tides. As the moon completes its monthly orbit around the Earth, it rises about 50 minutes later each day. Not coincidentally, the times of high and low tide also shift by a correspondingly predictable interval from one day to the next. At most locations, high and low tides occur 6 hours apart.

The Moon

Earth's satellite, the moon, is our nearest neighbor in space, at a distance of approximately 247,000 miles or 411,000 km. The moon is approximately one-quarter the diameter of Earth and has a mass estimated at 7.35×10^{22} kg. Although the Earth is over 80 times more massive than the moon, their mutual gravitational attraction accounts for most of the tidal bulge in Earth's oceans.

The moon travels around the Earth in an orbit that lasts an average of 27 days (the sidereal orbit). Relative to its position against the background stars, however, the moon's orbit around the Earth appears to last 29 days. This is known as the synodic orbit and is used as the basis of lunar calendars throughout the world.

The Sun

The moon is not the only celestial body whose gravitational attraction affects tides on Earth. Although the sun is located 93 million miles from Earth, our star is gigantic, representing 99% of the entire solar system's mass. Because Earth is held in orbit by the sun's gravity, it stands to reason that this gravitational pull affects Earth's tides. Although the sun's gravitational effect on Earth's tides is less pronounced than that of the moon, this effect manifests itself every two weeks as a phenomenon called spring tide.

Spring Tide

This phenomenon occurs twice a month at the new moon and full moon. High tides are higher than usual (and low tides are lower) because the sun, Earth, and moon are roughly in alignment at these times. As a result of this linear arrangement, the gravity of the sun and moon produces an additive effect on Earth's tidal bulge.

Neap Tide

Neap tides occur in the first quarter and last quarter of the moon. At these times, the sun, earth, and moon are located at right angles to each other. The result is that the difference between high and low tide is less pronounced at neap time than at any other time of the month.

Essay 12: Ursa Major Notable Features of the Constellation

Ursa Major, or the Great Bear, is one of the most widely known of all the constellations. One reason is that Ursa Major is the third largest constellation in the sky and very difficult not to notice. A second reason is the seven brightest stars of Ursa Major form an asterism, or pattern, called the Big Dipper. The Big Dipper, in turn, points the way to Polaris, the North Star. Many cultures around the world, from the ancient Greeks to the Native Americans, saw this star pattern as a bear. This article will explore some of the highlights of this constellation, focusing on its brightest stars and deep-sky objects.

For observers who live in the Northern Hemisphere, Ursa Major is visible on all clear nights of the year. From latitude 50 degrees N and northward, the stars of the Big Dipper are all circumpolar, meaning they never set below the horizon. As you can see from the diagram, the Big Dipper moves in a counterclockwise circle around the north pole of the sky. During winter, the Dipper is located in the northeast, standing upright relative to the northern horizon. In springtime, the Dipper is high overhead. Throughout summer, the Dipper is in the northwestern sky with its bowl pointed downward. For most of autumn, the Dipper can be found low in the northern sky just above the horizon.

Stars

In most constellations, the brightest star is designated alpha, the second brightest beta, and so on. In the case of Ursa Major, however, its stars were assigned Greek letters starting from Dubhe (alpha) and Merak (beta) in the Dipper's bowl and running to Alkaid (eta) at the end of the handle.

A line drawn through Dubhe and Merak points the way to Polaris, a second-magnitude star located about one degree from the north celestial pole. Continuing around the bowl, you come to the gamma and delta stars, named Phecda and Megrez, respectively. The epsilon star at the base of the handle is named Alioth, meaning 'the goats' in Arabic.

The zeta star is named Mizar, which means loincloth or wrapping. Close to Mizar is a dimmer star called Alcor (the rider). Mizar and Alcor are a naked-eye double. Observers with sharp vision really can distinguish two stars here.

In the medieval Arab world, the ability to see Mizar and Alcor distinctly served as a test of visual acuity. In Native American folklore, Mizar represented one of a trio of hunters pursuing the bear, and Alcor was his cooking pot.

The eta star at the end of the Dipper's handle is called Alkaid or Benetnash. In Arabian mythology, this star represents a chief mourner who leads his followers in a perpetual circle around the star Al Jadi (Polaris), seeking revenge for the murders of the children of a character named Na'ash. The name Benetnash translates to children of Na'ash.

Deep Sky Objects

Deep sky objects are often given an M designation in honor of the 18th-century French astronomer Charles Messier. He and his students compiled a list of objects, now known as galaxies, nebulae, and star clusters, frequently mistaken for comets by amateur astronomers. All of the Messier objects in Ursa Major are galaxies, with the exception of M97, the Owl Nebula. M81 and M101 are also 8th magnitude spiral galaxies, while M82, the Cigar Galaxy, is an irregular galaxy located near M81. M108 and M109 are dim 10th-magnitude spiral galaxies, best seen through a larger telescope. M51, in the adjacent constellation Canes Venatici (the Hunting Dogs), is called the Whirlpool galaxy. At magnitude 8, M51 is visible in a large pair of binoculars or any small telescope.

Biology & Ecology

Essay 13: Bioluminescent Fungi

The term bioluminescence refers to any source of light generated by living organisms. For the purposes of this article, the light in question is the visible spectrum or wavelengths from 400 to 800 nm. Most people are familiar with animal species capable of generating light, especially fireflies and certain aquatic species like deep sea fish. Scientists think the main functions of bioluminescence are attracting mates and prey as well as confusing predators with a sudden flash of light. As it turns out, the animal kingdom is not the only domain of bioluminescent organisms. Several species of bacteria, algae, and fungi are capable of bioluminescence as well.

In the realm of fungi, some species emit a faint glow, visible on dark, moonless nights. The fungal species most familiar to science is the foxfire, sometimes called will o' the wisp or fairy fire. This particular fungus is a saprophyte that grows on rotting tree bark. It is particularly noticeable in the fall as deciduous trees shed their leaves.

Scientists are unsure as to why some species of fungus are bioluminescent. One possibility is that the faint glow attracts insects, which proceed to disperse fungal spores. Thus, in the case of fungi, bioluminescence may serve as a form of reproductive assistance.

Light-generating reactions

Although many variations exist, photoemission tends to follow this basic pattern, sometimes described as a reversal of the oxygen-generating step in photosynthesis:

Luciferin + Oxygen + ATP ---> Oxyluciferin + ADP + Pi + Water + Light

In almost all bioluminescent organisms, the generation of light depends on an enzyme called luciferase. This enzyme catalyzes a series of reactions that utilize a molecular substrate (luciferin) in an unoxidized form, a source of oxygen, and, in most cases, adenosine triphosphate (ATP).

ATP supplies the free energy that drives the oxidation reaction forward. The enzyme hydrolyzes ATP, using this energy to combine oxygen with luciferin. When molecular oxygen binds to this substrate, photons are released. Some bioluminescent jellyfish contain a protein called aequorin, which emits light when calcium ions bind to it.

Other bioluminescent fungi

Over 40 species of bioluminescent fungi have been identified, mostly in tropical regions of the world. The main reason for this geographic distribution may be the abundant insect species in the tropics. Attracting a wider variety of insects may, in turn, lead to more effective spore dispersal. Scattering of spores remains the most plausible theory for fungal bioluminescence. After all, most fungi are seldom eaten, and spore formation usually occurs asexually.

Essay 14: Copper Toxicity in Sheep

The hallmarks of copper toxicity in mammals are anemia, lethargy, and liver failure. In humans, a rare disorder of copper metabolism called Wilson's disease results in chronic copper overload, which manifests as hepatolenticular degeneration. Over the course of several years, copper gradually deposits in the liver and basal ganglia of the brain leading to liver failure, motor deficits, impaired speech, and in some cases psychosis. Copper also accumulates in Descemet's membrane of the cornea; these greenish-gold deposits are known as Kayser-Fleischer rings. Acute copper poisoning in humans is rare. Symptoms include those seen in Wilson's disease as well as nausea, vomiting, and hemolytic anemia.

In the case of animals, sheep appear to be much more vulnerable to copper toxicity than cows, goats, or pigs. As with humans, chronic copper toxicity occurs far more often than acute poisoning. Intoxicated sheep exhibit extreme weakness, constant tooth grinding, and unquenchable thirst. When the animal's liver can store no more copper, the organ fails, releasing large amounts of copper into the bloodstream. This causes the destruction of massive numbers of red blood cells, in turn leading to kidney failure and death 1 to 2 days later.

Pigs and chickens can tolerate 10 times more dietary copper than can sheep. Hence, it is quite conceivable that soil containing significant levels of copper could be lethal to sheep or lambs while other livestock remain healthy. In addition to soil sample testing, a necropsy (animal autopsy) should be performed to identify copper deposits in the sheep's eyes, brain, and liver as direct evidence of toxicity. Copper levels in the animal's blood and liver would also be measured. Blood copper concentrations above 2 micrograms/mL (2 parts per million) and liver copper concentrations above 150 ppm are consistent with copper poisoning.

Soil contains an average of 30 to 50 ppm of copper; however, the actual concentration may vary from 2 to 250 ppm under natural conditions and reach far higher values (17,000 ppm) near copper mines and other industrial sites contaminated with metallic waste. Additional factors to consider are soil pH above 7.5 as well as the presence of clay, organic matter, and trace elements like molybdenum which bind to soil copper and prevent its uptake by plant roots.

During certain stages of growth, plants may be a more sensitive indicator of soil copper levels than the soil itself. At internal concentrations above 50 ppm, many plants display visible signs of copper toxicity, especially wilting or discoloration, including alfalfa, corn, oats, potatoes, soybeans, and wheat.

With the above in mind, soil samples with copper levels in excess of 61 ppm provide strong evidence that the sheep on a particular farm were exposed to soil containing high levels of the element. One likely scenario involves sheep feeding on grass, clover, alfalfa, or other plants containing significant amounts of copper. Other sources of environmental copper must then be ruled out, starting with animal feed and drinking water; these are the two most obvious routes of copper ingestion.

Other possibilities include copper fortified vitamin or mineral supplements intended for pigs or horses but mistakenly fed to the sheep as well as sheep grazing on pastureland fertilized with chicken or pig manure. Finally, liquid runoff from cattle footbaths often contains levels of copper toxic to sheep.

Treatment options for copper toxicity are limited. Molybdenum, manganese and vitamin C supplements added to feed often reduce excess copper buildup in livestock. In humans, a drug called penicillamine is used as a copper chelator.

Essay 15: Eukaryotes - Plant and Animal Cells

Both plant and animal cells are eukaryotic, meaning their genetic material is enclosed within a nuclear envelope. They also share many organelles in common, including a cell membrane, mitochondria, ribosomes, endoplasmic reticulum, lysosomes, and Golgi apparatus. Both animal and plant cells are aerobic, meaning they depend on oxygen to break down fuel molecules into the energy carrier ATP (adenosine triphosphate).

The major differences between animal and plant cells include the presence of a cell wall, chloroplasts, and large vacuoles in plant cells but not in animal cells. One organelle present in animal cells but absent in virtually all plant cells is the centriole. Each difference will now be discussed in detail.

Cell Wall

Plant cells share this feature in common with many species of fungi and bacteria. The cell wall forms a rigid envelope around the cell, but small molecules (water, sugars, and amino acids) can still diffuse across it. In plant cells, the cell wall is composed of cellulose, a polymer of glucose better known as wood. Fungal cell walls contain chitin (a polymer of glucosamine also found in crustacean shells). Bacterial cell walls are made of a variety of carbohydrate or protein polymers.

Chloroplasts

This organelle contains chlorophyll, a pigment that allows plant cells to perform photosynthesis; no animal cell is capable of this feat. Briefly, photosynthesis is a series of chemical reactions in which the energy of sunlight is harnessed to convert water and carbon dioxide into sugar. During photosynthesis, oxygen is released into the atmosphere. Consequently, all aerobic organisms depend on photosynthesis for their survival.

Chloroplasts contain their own DNA and ribosomes and replicate independently of the cell's nucleus.

In this respect, chloroplasts are similar to mitochondria. According to the Margulis hypothesis, both organelles originated as free-living prokaryotes, engulfed by primitive eukaryotic cells over 1 billion years ago. Since chloroplasts are confined to plant cells and certain species of protozoa, this event must have happened after plant and animal cells diverged evolutionarily.

Vacuoles

In plant cells, vacuoles are used to store sugar, starch, water, or waste. Vacuoles sometimes occupy a large part of a plant cell's interior. Some specialized animal cells, such as macrophages, phagocytose bacteria into a vacuole-like structure and then fuse this structure with a lysosome to destroy the bacteria. Other animal cells have small vacuoles or may lack vacuoles entirely.

Centrioles

These organelles are found in animal cells but are virtually absent in plant cells. Centrioles are composed of triplet arrays of microtubules arranged in a cylinder and linked by dynein arms. In animal cells, each centriole replicates prior to mitotic cell division. The daughter centrioles move to opposite poles of the cell and act as so-called microtubule organizing centers (MTOC).

Microtubules, which are cytoskeletal proteins made of tubulin polymers, align to form the spindle fiber during the prophase of mitosis. One end of each microtubule appears anchored near the centriole, while the other end attaches to the kinetochore, a structure located on a chromosome's centromere. When chromosomes separate after metaphase, each one moves along its respective set of microtubules toward the MTOC.

Plant cells also form a mitotic spindle made of microtubules, but their MTOCs consist of a dense area of poorly characterized proteins and generally lack centrioles. Since plant cells undergo mitosis as efficiently as their animal cell counterparts, it appears that centrioles are not necessary for eukaryotic cell division. Their disappearance from plant cells may reflect the ongoing divergence of animals and plants over long spans of evolutionary time.

How did Eukaryotic Cells Arise in the First Place?

Eukaryotic cells differ from prokaryotes in several ways, the most notable being a membrane-bound nucleus and the presence of complex organelles in the form of mitochondria and chloroplasts. Although it remains uncertain how eukaryotes evolved a nuclear envelope, the most straightforward mechanism would involve the internal folding of the cell membrane to surround free-floating DNA or chromosomes. Some bacteria exhibit the rudiments of nucleus formation in that their circular chromosomes are tethered to the cell membrane.

In the case of mitochondria and chloroplasts, evidence favors the endosymbiont theory, formerly called the Margulis hypothesis. This idea states that free-living aerobic bacteria were engulfed by larger, primordial cells that proceeded to trap the bacteria but not digest them. Over time, these bacterial prisoners lost most of their genes (which were transferred to nuclear DNA) except for those essential for replication. This accounts for the origin of mitochondria – these organelles require oxygen to operate the Krebs Cycle and Electron Transport Chain; a lipid bilayer surrounds them; they replicate independently (as opposed to the cell constructing them de novo); and they contain their own DNA, RNA, and prokaryotic ribosomes. Today, virtually all animal cells and fungi contain mitochondria and, not coincidentally, require oxygen to survive.

The argument for chloroplasts is similar. The primordial ancestors of protists and plants engulfed aerobic photosynthetic bacteria. Many of the bacterial genes were transferred to the nucleus; however, the chloroplasts still retained DNA, ribosomes, and, most importantly, the photosynthetic apparatus in their thylakoid membranes that generate oxygen. In this way, anaerobic eukaryotes became dependent upon oxygen for basic metabolic activities.

Essay 16: Fish Anatomy

On a superficial level, all species of fish are aquatic vertebrates with a head, body, and tail. For classification purposes, it is useful to divide fish into three main groups: jawless, cartilaginous, and bony.

Jawless fish are considered the oldest species of fish from an evolutionary standpoint. The two main varieties of jawless fish are hagfish and lampreys. Both are essentially parasites, latching on to other sea creatures and sucking nutrients from them directly.

Cartilaginous fish include sharks, skates, and rays. Their skeletons are composed entirely of cartilage; however, sharks are able to regenerate lost teeth. Many sharks shed tens of thousands of teeth over their lifetimes. Unlike sharks, rays and skates do not depend on teeth in order to feed but rather on bony plates that serve the same function. Many species of rays contain a venomous barb on their tails, which they use to paralyze prey.

Another interesting point of shark anatomy is the presence of a large, oil-filled liver, which helps keep the shark submerged. As almost all species of sharks live in salt water (which is more buoyant than freshwater), remaining submerged requires more adaptations compared to freshwater habitat. A further adaptation observed in sharks and other saltwater fish is their ability to concentrate salt in their tissues in order to maintain an osmotic balance with their environment. This metabolic process is not carried out by kidneys, which fish generally lack, but instead by specialized intestinal cells called the spiral valve.

Bony fish comprise the majority of all fish species. Their general characteristics include eyes, a rudimentary brain, a bony skeleton, a two-chambered heart, gills for gas exchange, and an extensive digestive tract. Most fish release ammonia directly into the water through their gills and, therefore, have no biological need for a urea cycle or kidneys. In most fish, reproduction occurs by means of external fertilization. Some fish build nests, and sea horses carry their eggs until they hatch, but few fish care for their young afterward.

When it comes to hunting and defense, some species of tropical fish produce venom, for example, scorpion fish and lionfish. These venoms usually act as toxins that paralyze or kill their target. Other species, most notably eels, have evolved jelly-like electric organs filled with specialized cells called electroplaxes. These cells are arranged in series like rows of miniature batteries; when these excitable cells discharge, their voltages are additive. Although an individual electro plaque cell may generate a voltage of only 0.15 volts, 5000 of these cells arranged in series can produce a discharge of 750 volts - enough to kill a person.

Essay 17: Fluorescent Tagged Proteins

Proteins tagged with fluorescent markers are extremely versatile cell and molecular biology research tools. As the name suggests, a fluorescent marker is a molecule that emits light of a particular wavelength following exposure to shorter wavelength photons, for example, a laser beam. If a fluorescent marker is of reasonable size and chemically stable enough to be attached to a protein, it is potentially useful to molecular biologists.

In many cases, the proteins in question are monoclonal antibodies generated against a particular target of interest, such as a cell surface receptor, a cytoskeletal protein, or a cellular organelle. In some cases, a fluorescent label is used to track the fate of a particular protein as it is processed in the ER and Golgi complex, transported along microtubules, or secreted from the cell altogether. In other cases, two or more fluorescently tagged proteins are introduced into the same cell to see whether they localize in the same or separate compartments or organelles.

One of the first techniques that made widespread use of fluorescent-tagged proteins was flow cytometry or FACS (Fluorescence Activated Cell Sorting). In this method, cells are incubated with antibodies conjugated with fluorescent dyes such as FITC (fluorescein isothiocyanate), which emits green wavelengths of light; PE (phycoerythrin) or Texas Red, both of which emit red wavelengths of light. Newer fluorescent dyes emit at yellow and blue wavelengths as well.

Following a 30 to 60-minute incubation, the cells are suspended in solution and aspirated into the flow cytometer. Once inside, the cells are channeled into a single file line and exposed to one or more laser beams. The machine then records the light signals emitted from the fluorescent antibodies attached to the cells. Depending on how many wavelengths the machine can detect, six or more cell surface receptors can be analyzed at once. Other fluorescence imaging techniques that have gained popularity over the past two decades include immunohistochemistry and fluorescence microscopy.

In addition to fluorescent dyes, fluorescent proteins (the entire protein acts as the marker) have come into widespread use in the last several years. Transgenic organisms expressing GFP (Green Fluorescent Protein) revolutionized the field of developmental biology starting in the 1990s. Simply by injecting the GFP gene into embryonic cells, scientists could now track the developmental fate of individual cells in organisms ranging from worms like C. elegans to vertebrates like zebrafish. Red and yellow fluorescent proteins (RFP and YFP) were developed shortly afterward; both are commonly used to visualize transgenic cells in vitro and in vivo.

Essay 18: Four Classes of Biological Macromolecules

In biology, a macromolecule refers to any agglomerate or polymer made up of smaller building blocks or monomers. Traditionally, the study of biological macromolecules was highly compartmentalized. While some scientists devote their entire careers to studying specific proteins, carbohydrates, lipids, or nucleic acids, the advent of the Human Genome Project and widely accessible online databases in the 1990s changed all of that. Today, the walls separating biochemistry, cell biology, genetics, and even biophysics have largely crumbled. This article briefly overviews each macromolecule and describes its major roles in cell biology and human physiology.

Proteins

In humans and other organisms, proteins are composed of 20 kinds of amino acids. Humans must obtain half of these amino acids from the diet, including phenylalanine, valine, threonine, tryptophan, isoleucine, methionine, histidine, arginine, leucine, and lysine. The first letter of each of these so-called essential amino acids forms the mnemonic "PVT TIM HALL."

The sequence of amino acids specifies the primary structure of a certain polypeptide (or protein). Proteins range from about a dozen to several thousand amino acids in length. Because the number of potential amino acid combinations is determined by 20 to the power of n, the variety of proteins that can be produced is nearly limitless.

Proteins are the foundation of structure and function in all known cells. This is also true on the tissue and organ levels. A few examples of structural proteins include actin microfilaments and microtubules at the cellular level. At the macroscopic level, myosin is a key component of muscle fibers, and collagen serves as the scaffold in skin, teeth, and bones.

Examples of purely functional proteins include the enzymes that catalyze thousands of chemical reactions in our bodies; hemoglobin to transport oxygen in the bloodstream; neurotransmitters such as endorphins; protein hormones such as insulin; ion channel pumps to conduct nerve impulses; and immunoglobulins, a.k.a. antibodies, which are a major branch of the specific immune response.

Carbohydrates

More kinds of sugars exist than any other type of biological polymer. The two most common polysaccharide polymers are cellulose, long rigid strands of glucose more familiar to us as wood, and chitin, a polymer of N-acetylglucosamine found in the shells of crustaceans.

In humans, most carbohydrates are either metabolized for energy or converted to fat for long-term energy storage. Some glucose is converted to starch, or glycogen, a branched polymer found in the liver and skeletal muscles. The body stores enough glycogen to meet its glucose needs for approximately 24 hours.

Some carbohydrates are attached to proteins. Scientists think this accomplishes two things. First, sugars may extend the life span of a given protein by blocking its degradation by protease enzymes. Second (arguably more important), sugars act as intracellular address labels, directing the protein to its appropriate destination (e.g., the cell nucleus, mitochondria, plasma membrane, or other organelle).

Because humans can synthesize glucose from amino acid precursors and, in turn, convert glucose into other sugars (fructose, mannose, ribose), there is no such thing as an 'essential sugar' in human physiology.

Lipids

Fat is the first image that comes to mind when the word' lipid' is mentioned. Although this is true, lipids encompass a variety of other molecules, including steroid hormones and bile acids, both derived from cholesterol, as well as signaling molecules such as inositol triphosphate, prostaglandins, thromboxanes, and leukotrienes.

From a macromolecular perspective, most lipids fall into three categories: triacylglycerols stored in adipocytes (fat cells); diacylglycerols found in all cell membranes; and sphingomyelin, an abundant component of the myelin sheath surrounding neuronal axons throughout the human nervous system. Adipose tissue serves as the main energy reservoir in periods of starvation. Gram for gram, fatty acids yield nearly three times as much energy in the form of ATP than do proteins or carbohydrates. As with amino acids, certain fatty acids (e.g., omega-3 fatty acids) are considered essential because humans must obtain them from their diet.

Nucleic Acids

DNA (Deoxyribonucleic Acid)

Nowadays, it is common knowledge that genes are made of DNA. Surprisingly, it took biologists the first half of the 20th century to arrive at this conclusion. Until the 1920s, relatively little was known about the components of DNA or the central role DNA plays in determining hereditary traits. Building on the work of several scientists, James Watson and Francis Crick proposed their double helical model of DNA in 1953.

In the Watson-Crick model, each strand of DNA is composed of a backbone formed of deoxyribose sugars linked together by phosphodiester bonds. A nitrogenous base is attached to each sugar – either a two-ringed purine, adenine (A), or guanine (G) or a one-ringed pyrimidine - thymine (T) or cytosine (C). Each base pairs with a complimentary base: A pairs with T, and G pairs with C. Two complimentary DNA strands wrap around each other to form a structure called an anti-parallel double helix. The two DNA strands are held together by numerous hydrogen bonds that form between their respective base pairs.

A decade after Watson and Crick's discovery, three teams of scientists cracked the genetic code. Essentially, they discovered that DNA is read as a non-overlapping triplet code in almost all organisms, in which each unit of three bases corresponds to one amino acid.

This unit (upon being transcribed into messenger RNA) is called a codon. Except for tryptophan and methionine, all other amino acids are specified by anywhere from two to six different codons. Three codons (UAG, UGA, and UGG) do not code for amino acids. Instead, they serve as stop signals that terminate protein translation.

RNA (Ribonucleic Acid)

RNA is similar to DNA in some respects. It is composed of a sugar-phosphate backbone and contains three of the same bases as DNA (adenine, guanine, and cytosine). Instead of thymine, however, RNA contains uracil, which is chemically identical to thymine except that it lacks a methyl group. Another difference is that DNA remains confined to the nucleus (in eukaryotic cells), whereas RNA is transcribed in the nucleus and then departs into the cytosol. Finally, in contrast to DNA, cells contain several distinct types of RNA, including messenger RNA (mRNA), transfer RNA (tRNA), ribosomal RNA (rRNA), and, in eukaryotes, small nuclear RNA (snRNA).

Regarding their roles, mRNA carries the message encoded in DNA to the ribosome for translation into protein. Transfer RNA acts as a tow truck hauling each amino acid to the ribosome. Ribosomes themselves are large complexes of peptides and ribosomal RNA that assemble amino acids into new proteins. Finally, snRNA is confined to the nucleus and helps process mRNA into its mature form.

Many scientists are convinced that primitive cells used RNA as their genetic material before DNA became the dominant nucleic acid. Two lines of evidence support this theory: 1) Certain viruses, including influenza and HIV, contain genomes made entirely of RNA. Viruses themselves may have originated as rogue genetic elements that broke away from their parent cells and somehow acquired the ability to infect other cells. 2) Unlike DNA, certain types of RNA have catalytic abilities. Small nuclear RNA splices mRNA by removing sequences called introns, then rejoins the remaining exons before the mature mRNA is allowed to exit the nucleus. Ribosomal RNA catalyzes the formation of peptide bonds, performing the fundamental step of protein synthesis. Scientists sometimes call RNA-based catalysts "ribozymes" in contrast to enzymes, which are made of protein.

Essay 19: Honeybees

Many people are surprised to learn that the European honey bees introduced into North America after the year 1600 were not the first bees domesticated in the New World. It turns out the Maya of Central America domesticated a species of stingless bee for honey production centuries before the arrival of the Europeans. Some species of stingless bees can still be found in Mexico's Yucatan peninsula; however, as is often the case with invasive species, European honey bees have replaced many of the native bee species in North America.

In the lands north of Mexico, European settlers imported honey bees primarily to produce honey and later to pollinate crops. As English settlers spread westward, honeybees spread with them. In fact, the appearance of honey bees often preceded the arrival of Europeans themselves. Native American tribes soon came to regard honey bees (which they nicknamed "the white man's fly") as a bad omen.

By the 20th century, apiaries grew into a gigantic industry, as farmers and owners of large orchards, especially those containing apple and cherry trees, came to rely on huge numbers of honey bees to pollinate their crops. In the short term, this led to an explosion in the honey bee population. The downside was that honey bees became a genetically inbred species even as their total numbers grew.

The reason for this seemingly paradoxical event involves the reproductive life of bees. As with other species of social insects (e.g., ants and termites), the worker bees are all females. They hatch from eggs laid by a queen bee, which must mate only once in her lifetime (which lasts 3 to 5 years in contrast to a few months for most workers). The male bees, or drones, arise from unfertilized eggs laid by the queen or workers.

At any rate, industrial apiaries bred their queens in virtual isolation from other colonies over the years. Predictably, this has made most industrially raised honey bees vulnerable to fungal infections, mites, and probably viruses as well. Over the past five years, billions of honey bees have vanished due to an as yet unidentified disease termed colony collapse disorder.

Another bee experiment that went woefully awry was a breeding project begun in the 1950s in Brazil. Scientists believed that crossing African bees and European honey bees would produce an ideal hybrid bee that would be docile, withstand tropical heat and pests, and produce abundant stores of honey. Unfortunately, the hybrid offspring, the so-called killer bees, were extremely aggressive and produced little honey; however, they were hardy to the point of escaping from Brazil and spreading all the way to the southern United States.

Essay 20: LTP and Synaptic Transmission

The term long-term potentiation, or LTP, refers to a strengthening of synaptic connections between two neurons following intense, repetitive stimulation of the presynaptic neuron. Most neuroscientists think this process plays a fundamental role in learning and memory at the cellular level. This article will offer some historical perspective on the discovery of LTP in the context of neuroscience's ongoing quest to answer a profound question: how does the brain learn and remember information?

The idea that the brain is responsible for thoughts and memory goes back to ancient times, appearing in the works of Hippocrates (4th century BCE). For the next 23 centuries, the mechanisms underlying learning and memory remained unknown; nevertheless, a general consensus emerged that the brain could somehow perceive, respond, and adapt to environmental stimuli. By the 19th century, neuroscientists proposed that electrical, chemical, and possibly structural changes in the brain accompanied learning and memory. That was about as far as neuroscience would progress until the mid-20th century. Even at this late date, many scientists resisted the idea that an adult's brain's structure could change in response to novel (or repetitive) stimuli.

In 1949, the paradigm began to shift after Donald Hebb proposed that synaptic connections within a neural network become strengthened when the neurons fire frequently. Conversely, unused or seldom-used synaptic connections tend to weaken over time and may disappear entirely. Hebb's hypothesis is often expressed in the quote, "Cells that fire together wire together."

Although Hebb's emphasis on synaptic plasticity was slow to gain acceptance, in retrospect, it marked the dawn of a new era in neuroscience. The next major breakthrough came in 1968 when Terje Lomo and Tim Bliss demonstrated LTP induction in the hippocampus, a region of the brain's temporal lobes involved in the consolidation of long-term memories.

The search for the retrograde messenger

Although the early 1970s understood the basic sequence of LTP, many puzzles remain unsolved. Chief among these is an adequate explanation for the changes in the presynaptic neuron underlying its enhanced ability to release neurotransmitter (NT). In other words, scientists agree that a hallmark event in LTP is increased presynaptic NT release, but the molecular mechanisms remain obscure.

By the early 1990s, several theories had been proposed; nearly all of them invoked retrograde signaling from the postsynaptic neuron to the presynaptic cell. Some scientists focused on the excitatory amino acid glutamate along with calcium ions. Supposedly, after the presynaptic neuron fired a sufficient number of action potentials, the backwash of one or both substances onto the presynaptic cell induced LTP.

Other scientists turned their attention to smaller, more exotic molecules, including soluble gases like nitric oxide or carbon monoxide, the membrane lipid arachidonic acid, the energy carrier ATP (adenosine triphosphate), PAF (Platelet Activating Factor), and others. Predictably, the debate generated more heat than it did light, and at present, no consensus exists as to the molecular mechanisms responsible for the induction phase of LTP.

New Directions

In recent years, some researchers have focused on the long-term preservation of LTP, with a particular emphasis on brain-derived neurotrophic factor (BDNF). This protein, a nerve growth factor family member, seems to play many important roles in the CNS, including neuronal survival. In regards to LTP, BDNF may play a presynaptic role, enabling neurons to sprout new axonal connections and enlarge existing synaptic terminals. BDNF may also promote dendrite outgrowth on postsynaptic neurons, although a host of other protein growth factors are likely to play important roles as well.

Essay 21: Macronutrients, Dangerous Diets, and Starvation

Macronutrients are the basic foundation of the human diet in childhood as well as adulthood. This article will focus on the macronutrient requirements of growing children, which fall into three categories: carbohydrates, proteins, and fats/lipidsd.

Carbohydrates

These can be roughly divided into two groups: simple sugars, such as table sugar, corn syrup, and honey, and complex carbohydrates, including whole grains, potatoes, rice, and other forms of starch. As a general guideline, carbohydrates should provide up to half of a growing child's daily caloric needs, which works out to 1,000 or so calories derived from carbohydrates.

Proteins

These are found in meat, seafood, poultry, dairy products, and legumes. Gram for gram, proteins contain approximately the same number of calories as carbohydrates (4 kcal/g). Pound for pound, growing children have a higher protein requirement than adults. Proteins should constitute a minimum of 25% of a child's daily caloric intake, or around 500 calories.

Proteins are the source of amino acids, the main source of usable nitrogen for humans and other mammals. Certain amino acids are referred to as essential amino acids because the human body cannot synthesize them; therefore, they must be obtained from the diet.

In children, these ten amino acids are phenylalanine, valine, threonine, tryptophan, isoleucine, methionine, histidine, arginine, leucine, and lysine. Adults can synthesize enough arginine in the urea cycle to compensate for any dietary deficiency. A mnemonic for remembering the essential amino acids is PVT TIM HALL.

A deficiency of even a single essential amino acid can put a child into a negative nitrogen balance. This manifests itself first as a failure to gain weight, followed by a failure to grow in stature. Protein malnutrition in the context of sufficient calories is called kwashiorkor, a West African word that translates to "disease of the displaced child." Children whose diet consists of a single source of calories, such as corn, rice, or millet (individually, not in combination), are vulnerable to developing this form of protein malnutrition.

Lipids

These include animal fats as well as plant-derived fats and oils. Gram for gram, fats contain over twice the number of calories as proteins or carbohydrates, approximately 9 kcal/g. Fats should make up 25-30% of a child's daily caloric intake.

Certain fatty acids are essential and can only be obtained from the diet. In humans, these include linoleic and linolenic acids, which are found in a variety of sources, including olive, sunflower, and flax seed oil. Omega-3 fatty acids are also part of a healthy diet. Salmon, tuna, haddock, and other cold-water fish are rich sources of omega-3 fatty acids.

Micronutrients comprise dozens of vitamins and minerals. They are necessary for life but not as sources of energy.

Dangers of Low-Calorie Diets

Many potential dangers are associated with very low-calorie diets; the most common are ketosis (leading to full-blown ketoacidosis), electrolyte abnormalities, a negative nitrogen balance, and, in extreme cases, death due to heart, liver, or kidney failure.

Ketoacidosis

When caloric intake is insufficient to meet the body's energy requirements, the body shifts its metabolism into starvation mode. In doing so, the body is walking a tightrope, so to speak. On the one hand, the body tries to maintain blood glucose at or above 50 mg/dL. If glucose levels fall below 50, the effects are dramatic: tremors, sweating, seizures, and even coma occur. On the other hand, the body must figure out how to supply the brain and other vital organs with fuel in the absence of a reliable supply of calories from the diet.

The body's solution is to burn through its fat reserves. Although gram for gram, fat contains over twice as many calories as sugar or protein, a new problem arises: the human body can convert sugar into fat but cannot convert fat into sugar. As a professor of mine once put it, sweets can make you fat, but fats can't make you sweet.

In biochemical terms, fatty acids can only be burned aerobically by a process called beta-oxidation. This pathway yields abundant amounts of a molecule called acetyl CoA, which enters the Krebs cycle and electron transport chain to generate ATP. Unlike plant cells, mammalian cells cannot convert acetyl CoA to its immediate predecessor, pyruvate. (Unlike acetyl CoA, pyruvate can be converted to glucose in a series of enzymatic reactions known as gluconeogenesis).

In response to this biochemical limitation, the liver and kidneys slowly break down proteins into amino acids, most of which can be converted to pyruvate and ultimately into glucose.

This glucose is not burned for energy; however, most is used to maintain blood glucose levels above 50 mg/dL. The bulk of the body's energy requirements, especially the brain, comes from metabolites of fatty acid breakdown called ketone bodies.

Although this strategy allows the body to survive periods of starvation, some people, especially diabetics, are susceptible to out-of-control ketone body production, a potentially fatal state called ketoacidosis. The liver and kidneys cannot clear high levels of ketone bodies quickly enough. The ketone bodies accumulate in the blood, where they acidify blood pH, causing it to drop to an unsafe level (below 7.20). Essentially, the normal buffering systems in the bloodstream (bicarbonate ions and albumin) are overwhelmed by the acid load. As blood pH drops, the person begins to hyperventilate in an effort to rid the blood of CO_2.

Unless metabolic acidosis is treated promptly, vomiting, seizures, and coma ensue, as the brain simply cannot tolerate the high acid level.

Electrolyte Abnormalities

These occur in the context of ketoacidosis, in particular, a rise in blood potassium levels called hyperkalemia. In an effort to raise blood pH, body cells take in protons (H+) and release potassium ions (K+) in order to maintain electrical neutrality. Given enough time (several days), the kidneys can restore order by disposing of the excess H+ and K+ ions in the urine. Over the course of hours, however, the kidneys cannot correct the imbalance, and cardiac arrhythmias can result.

Negative Nitrogen Balance

The result of caloric restriction (especially in growing children) may be protein malnutrition. A deficiency of a single essential amino acid will tip the body into a metabolic state in which more protein is broken down than is synthesized. This state, known as a negative nitrogen balance, is marked by a rise in blood levels of nitrogen in the form of urea (in medical terms, called elevated BUN). In the presence of dehydration, liver, or kidney failure, an elevated BUN (azotemia) can lead to coma and death. Under less extreme circumstances, the body can usually tolerate moderately elevated BUN.

Starvation

Prolonged starvation (~ 90 days) culminates in death; the proximate causes include heart, liver, or kidney failure. Cardiac failure (in this case, due to atrophy of the myocardium) is a setup for fatal arrhythmias. Hepatic failure results in the collapse of albumin production, leading to severe ascites (leakage of fluid from the bloodstream into the abdominal cavity). Eventually, the urea cycle ceases to function, which leads to elevated blood ammonia levels. Hyperammonemia, as it is called, leads to hepatic encephalopathy, marked by seizures, coma, and death. Untreated renal failure leads to uremia (an acidotic state similar to ketoacidosis), which also culminates in coma and death.

Essay 22: Neuroscience Predictions

I think neuroscience will advance, but I think it will hit a brick wall when it comes to explaining consciousness and identity at the molecular level. In essence, the reductionist approach simply falls apart whenever we try to dissect phenomena at the systems level (such as how the brain generates ideas or self-awareness).

Biology faces a similar problem that physicists confront in their quest for a comprehensive theory uniting general relativity and quantum mechanics. At a fundamental level, the theory breaks down when we ask, "Why did the parameters of the theory turn out the way they did?" or "How did the 'rules' themselves arise?"

This is what I mean. In physics, the Standard Model predicts and describes all the fundamental particles observed in colliders. No explanations exist, however, for questions like "How can the electron have a charge equal and opposite to that of the proton, yet the electron is only 1/1836 as massive?"

The theory itself offers no insight because it is purely descriptive and takes the constants of nature at face value. When asked to explain why particles have the properties they do, we resort to shrugs and hand waving in the end.

Similarly, in neurobiology, we can describe in painstaking detail how particular ions move through their respective ion channels, stimulating or inhibiting the release of neurotransmitters from a given neuron. We also know, in rather good detail, how neural circuits are connected and which synapses are excitatory, inhibitory, or modulatory.

Yet our vast store of knowledge cannot explain how the brain understands how individual neurons work, how humans make decisions based on free will or the ultimate mystery, how the CNS is aware of its existence.

I won't say these mysteries are impossible for science to solve, but I do believe that the solutions will elude us for quite some time.

Essay 23: Rainforests of the World

Rainforests exist in both temperate and tropical regions of the world. Scientists generally consider a forest to be a tropical rainforest if it is located between the Tropics of Cancer and Capricorn, receives at least 80 inches of rain annually, and never experiences frost or freezing temperatures. In contrast, some forests of the Pacific Northwest are considered temperate rainforests, the main differences being extremes of temperature and geographic location. Unfortunately, in less than a century, humans have destroyed over half of the world's rainforests, both temperate and tropical.

Today, tropical rainforests cover a mere 2% of the Earth's surface but are home to over half of all animal and plant species on the planet. Rather than compile a list of rainforest animals, it makes a great deal more sense to explore each layer of the rainforest, starting with the emergent layer of the forest canopy. Afterward, animals unique to specific rainforests around the world will be discussed.

Emergent layer

The tallest trees in the rainforest grow to a height of nearly 200 feet. Many bird species, notably hummingbirds, macaws, parrots, and eagles, can be found here, along with bats and other flying insects. Tree snakes and small monkeys may be found here as well.

Canopy

This is the upper layer of trees, inhabited by a dizzying array of birds, snakes, amphibians (tree frogs), insects, sloths, and monkeys.

Understory

As with the canopy, a tremendous variety of animals live in the understory, including monkeys, jaguars, snakes, sloths, and, of course, insects, including some large species of butterfly.

Forest floor

A number of years ago, one ecologist claimed that more was known about the surface of the moon than about the floor of the tropical rainforest. While that may be an exaggeration, it is true that the animal species in this zone are more difficult to study than in most other parts of the forest. A thick cover of seedlings and other plants conceals nearly the entire forest floor from view. Surprisingly, little sunlight reaches the forest floor, with less than 5% of the light hitting the treetops.

Some better-known arthropod species include leaf-cutter ants, a huge variety of beetles, and numerous types of spiders, including some of the largest known to science. Annelids live here as well, including leeches, centipedes, giant millipedes, and their close kin, the worms. Frogs, toads, salamanders, lizards, snakes, and many rodent species spend at least part of their time on the forest floor.

No Two Rainforests Are Identical

In the rainforests of Central and South America, jaguars, coatis, giant anteaters, and tapirs roam the forest floor, although jaguars spend a great deal of time in trees. The cloud forests of Central America are home to the endangered quetzal bird, whose feathers were prized by the ancient Maya. The Amazon remains the largest of the world's tropical rainforests and boasts the world's largest snake, the anaconda. Tropical South America is also home to the world's largest rodent, the capybara.

Tropical rainforests once covered most of equatorial Africa, and the Congo basin still contains the second-largest rainforest after the Amazon. This forest is home to many unique animals, including okapis (which resemble a cross between a giraffe and a zebra), chimpanzees, and lowland gorillas.

The rainforest of Madagascar is home to many of the world's chameleons along with some species found nowhere else on earth, for example, lemurs, their relative the aye-aye, the Madagascar flying fox, and a cat-like creature called the fossa.

Southeast Asian rainforests, especially those of Borneo, Java, and Sumatra, are home to some critically endangered species, including tigers, orangutans, and two kinds of rhinoceros. Tigers also live in the mangrove swamps of India, but their total number has dwindled to some 3,200 in the wild, as opposed to an estimated 100,000 a century ago. Elephants also live in the shrinking forests of India, Thailand, and Laos. In 1992, a species of antelope unknown to Western science called the saola was discovered in a remote forest bordering Laos and Vietnam.

India's northeastern region of Assam has stretches of intact rainforest; small pockets of forest line India's coastal mountain ranges. The Andaman Islands, also part of India, contain some of the most pristine rainforests left in tropical South Asia. Some animals found in Indian rainforests include cloud leopards, elephants, hornbills, the Indian rhinoceros, macaques, and several other primate species.

One of the most remote and least explored rainforests is located on the island of New Guinea. No monkeys are native to this island; instead, creatures like tree kangaroos and flightless cassowaries have evolved to become the largest animals in New Guinea's rainforest. This island is also home to many species of birds of paradise (as well as their floral namesake); bower birds, which court females by building nests large enough for a child to fit in; and 43 other bird species found nowhere else on earth.

The flooded forest

This phenomenon occurs in parts of the Amazon for several months each year. The trees and other plants of the flooded forest somehow survive several months of partial or total submersion during the rainy season. Many of these plants depend on fish to swallow and disperse their seeds. Considering how a number of fish species have declined or disappeared entirely from parts of the Amazon, the repercussions for plant life are alarming. Many animals are unique to the Amazon River itself, including a species of pink dolphin, a gray dolphin, a manatee, South America's largest river turtle, and the black caiman. More species of fish are believed to live in the Amazon River than in the entire Atlantic Ocean.

Earth Science & Geology

Essay 24: Carbon-14 Dating

Until the late 1940s, archaeologists and anthropologists faced an uphill battle when it came to dating early modern human remains and other prehistoric artifacts. At the time, the most reliable technique consisted of a relative dating method devised by geologists known as stratigraphy. The basic premise of stratigraphy is that fossils and artifacts found between two undisturbed layers of rock must be younger than the rocks beneath them but older than the rocks deposited above them. Based on the rate of decay of uranium and other radioactive isotopes, a ballpark estimate could be obtained for the age of some rock formations and, by extension, the approximate age of fossils found within them.

Although stratigraphy was obviously better than random speculation, it was useless for dating fossils found at geologically recent sites, especially those less than 100,000 years old. In human terms, 100,000 years is a vast stretch of time; however, in the grand scheme of Earth's natural history, it represents the proverbial blink of an eye. Matters stood at an impasse until 1947 when Willard Libby developed a new dating technique based on the half-life of carbon 14, a radioactive isotope discovered in 1940 by Kamen and Reuben.

A brief explanation of isotopes and half-life is in order at this juncture. The vast majority of carbon atoms in the universe occur as carbon 12, meaning their nuclei contain 6 protons and 6 neutrons. From a nuclear standpoint, carbon 12 atoms are completely stable and do not undergo any known form of radioactive decay. This pattern holds for most atomic elements, except for massive nuclei like uranium, of which all isotopes are inherently radioactive.

Unlike stable isotopes, radioactive nuclei decay at a constant rate over time. That rate may be very rapid, on the order of hours for a synthetic element called Technetium-99, or unimaginably slow, for example, 4.5 billion years in the case of Uranium-238. A half-life is defined as the amount of time it takes for half of the atomic nuclei of a radioactive element to decay into some other form.

A crucial point to remember is that only half of a sample of a radioactive isotope decays during each successive half-life. To illustrate this point, suppose you start with 100 grams of a radioactive isotope whose half-life is 100 days. 50g of the isotope would remain after 100 days, 25g after 200 days, 12.5g after 300 days, 6.25g after 400 days, etc. After ten half-lives have elapsed, the amount of radioactivity would fall to a nearly undetectable level.

In our hypothetical scenario, this would be equal to 1,000 days. Scientists consider ten half-lives to be the maximum length of time for which a given radioactive isotope can yield an accurate date.

In the case of carbon, less than 1% of this element exists as carbon 14, which contains 6 protons and 8 neutrons in its nucleus. The two extra neutrons destabilize carbon 14 so that at some point in time, its nucleus decays by a process known as beta emission. In this form of radioactive decay, a neutron is converted into a proton as well as a high-energy electron, which gets ejected from the atom as a beta particle.

In this case, beta emission produces an atom of nitrogen 14, whose nucleus consists of 7 protons and 7 neutrons. Electrons have virtually no mass; only the mass of protons and neutrons are counted in radioactive decay equations.

Carbon 14 is an incredibly significant isotope for two reasons. First, carbon is ubiquitous in the cells of humans and all other organisms. Proteins, carbohydrates, lipids, and DNA are all carbon-based macromolecules. In fact, hydrogen and oxygen are the only elements present in humans in greater abundance than carbon (considering that almost 60% of human body weight consists of water). As such, a small amount of 14C is likely to be trapped inside a person's bones for centuries or millennia after that person's death.

Body parts other than bone may be preserved under certain circumstances, like burial sites in the desert or permanently frozen regions. As a general rule, however, only bones and teeth are likely to be found intact.

Second, the half-life of carbon 14 is estimated at 5,700 years, an ideal period for dating both ancient and prehistoric artifacts. Because the amount of 14C present in a sample becomes indistinguishable from levels of background radiation after ten half-lives have elapsed, objects can be dated accurately to a maximum age of 57,000 years. Except for prehistoric human sites in Africa (and possibly the Middle East), all other modern human remains, and by extension, human artifacts, fall within this age limit.

This knowledge has proved extremely valuable in dating prehistoric human sites outside of Africa, especially those in the Americas, which humans colonized toward the end of the last Ice Age 12,000 years ago. Artifacts from the ancient world, including Egyptian mummies and the Dead Sea Scrolls, have also undergone radiocarbon dating. As a result, their true ages have been established or corroborated to within a few centuries.

Essay 25: Climate Change over the Past Thousand Years

For the purposes of understanding global climate change, it is useful to divide the last millennium into three eras: the Medieval Warm Period, the Little Ice Age, and the Fossil Fuel Era, marked by the Industrial Revolution and global warming.

The Medieval Warm Period: 800 – 1300

Most theories invoke a combination of increased solar energy output and decreased volcanism to explain the relatively warm climate of the Middle Ages. These factors do correspond to several warm periods, according to dendrochronologists, who study tree rings, and climate researchers, who study ice core samples. Nevertheless, 500 years is the proverbial blink of an eye on the scale of geologic time. The debate is likely to continue for the foreseeable future.

Regardless of its cause(s), the Medieval Warm Period saw an increase in Europe's population, the Viking colonization of Iceland and Greenland, and the Inuit expansion from northern Canada to Greenland a few centuries later. In the American Southwest, the Anasazi culture of Chaco Canyon reached its peak around 1100. Polynesians reached Easter Island sometime after 800 AD in the South Pacific, and the Maori colonized New Zealand around 900 AD.

The Little Ice Age: 1300 – 1820

Compared to the medieval warm period preceding it, global temperatures became cooler and rainfall patterns more unpredictable starting around the year 1300. Theories abound as to the cause of these events, but most focus on one or more of the following factors:

1) An increased number of volcanic eruptions led to the reflection of more sunlight back into space, accounting for the colder temperatures. Although volcanic ash does block sunlight from reaching the Earth's surface, a significant increase in global volcanism is difficult to discern from the historical record. For instance, the eruption of Mt. Tambora in 1815 was estimated to be many times more powerful than Mt. St. Helens but occurred near the end of this cold period. By the time Krakatoa erupted near Java in 1883, the Little Ice Age had been over for several decades.

2) Decreased solar output. Sunspots (a marker of solar activity) wax and wane in a cycle lasting 11 years or so. During the 17th century, however, records indicate that sunspots practically vanished for years on end. No consensus exists as to the cause of this event, known as the Maunder minimum.

Interestingly, the coldest decades of the Little Ice Age also occurred during the mid-1600s, suggesting these two events were more than a mere coincidence.

3) Alterations in deep ocean currents. The Gulf Stream and Japanese currents move vast amounts of warm salt water from the equator toward the poles. This process, termed thermohaline circulation, contributes to a mild climate in places as far north as the British Isles and the Pacific Northwest of North America. Without the Gulf Stream, for example, the climate of England would become as frigid as that of the Hudson Bay in northern Canada. Something similar may have happened around the year 1300.

In one scenario, melting Arctic glaciers during the medieval warm period decreased the North Atlantic Ocean's salinity to the point that heat convection in the ocean depths practically shut down. The ensuing cold spell lasted for over 500 years.

The consequences of the Little Ice Age were largely negative for Europe and elsewhere. Cold, damp conditions favored the spread of rats that carried the bubonic plague to Europe in the mid-1300s. The same climate also promoted the growth of ergot fungus, which contaminated grain supplies in much of Northern Europe. Ergot poisoning plagued Europe for centuries. To superstitious people living in a pre-scientific era, the frightening, bizarre symptoms of ergot toxicity, most notably burning sensations in the extremities, miscarriages, and hallucinations, were clear evidence of witchcraft. Not surprisingly, hysteria over witches reached a fever pitch in Europe during the Little Ice Age.

Cultures in other ecologically sensitive places declined or collapsed altogether during the Little Ice Age. The Polynesian population on Easter Island plummeted by 90% in the wake of deforestation and severe drought. At Chaco Canyon, the Anasazi succumbed to a combination of drought, famine, and warfare. The Norse colony in Greenland vanished in the decades after 1410. In Cambodia, the Khmer capital of Angkor Wat was virtually abandoned after 1432. Decades of crop failure in China culminated with the Manchus overthrowing the Ming Dynasty in 1644.

The Fossil Fuel Era and Global Warming: 1820 – Present

Humans have burned small quantities of coal and oil for thousands of years, but that amount was minuscule compared to the massive fossil fuel consumption that powered the Industrial Revolution. As the graph in the reference section illustrates, the burning of coal, oil, and natural gas has accelerated since 1850 and shows no sign of slowing down today, at least not at the global level. An inevitable byproduct of burning carbon compounds is the production of far more atmospheric carbon dioxide than green plants, and phytoplankton can extract via photosynthesis. CO2, methane, and other greenhouse gas emissions have skyrocketed, especially after 1900. Annual greenhouse gas emissions are estimated at over 8 billion metric tons worldwide.

Not coincidentally, average global temperatures have increased by 1 or 2 degrees Celsius in that time span. Today, all major glaciers are in retreat. In addition, massive ice caps melting in Greenland and Antarctica point to a clear warming trend. Is global warming a human-driven phenomenon? Most environmental scientists think the answer is yes, considering the tremendous amount of fossil fuels burned in the last 50 years alone.

Even though world oil reserves are dwindling, coal and natural gas supply is expected to last for several more centuries. Even at this stage, however, the consequences of global warming are becoming apparent. They include more frequent droughts in the U.S., Australia, and parts of Africa, more intense hurricanes fed by warmer ocean waters, and a gradual rise in sea levels caused by melting glaciers and polar ice caps.

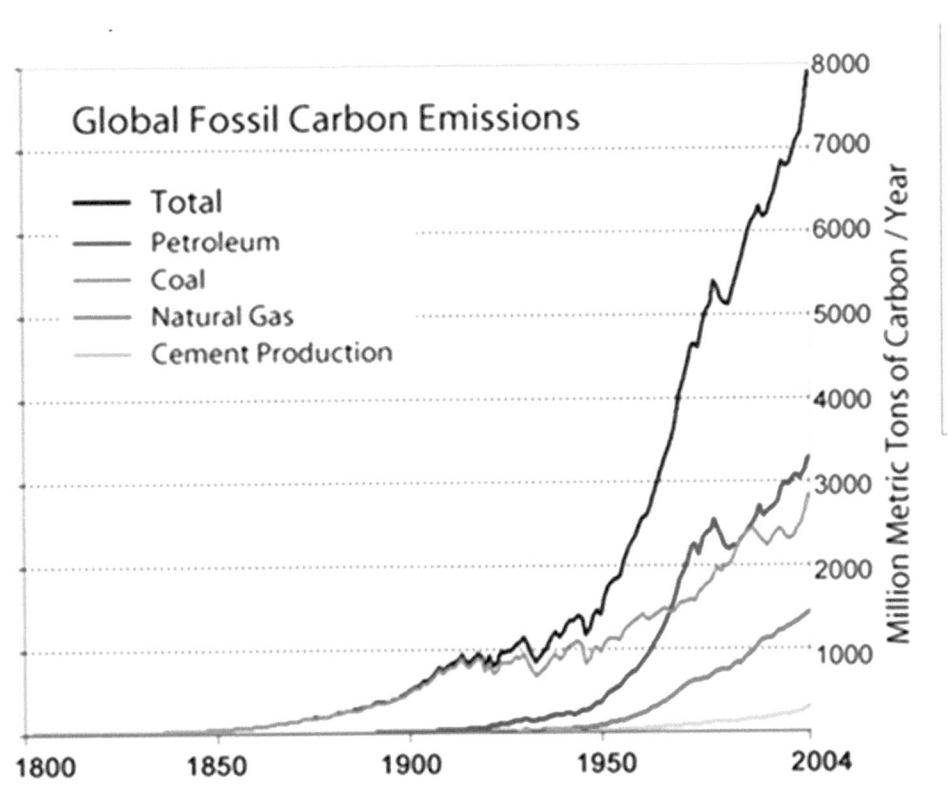

Essay 26: Geography of the Atlantic Ocean

The Atlantic Ocean is the second largest of the world's oceans, measuring over 41 million square miles and covering approximately 20% of the Earth's surface. The continents bordering the Atlantic include North America, South America, Europe, and Africa. The Atlantic Ocean basin is mostly surrounded by land. It extends from subarctic latitudes through the tropics all the way to the Southern Ocean surrounding Antarctica. This article will focus on the major geographical features of the Atlantic Ocean and their surprising impact on marine biology and global climate.

Continental Shelves

While the average depth of the Atlantic Ocean is over 12,000 feet (~ 3600 meters), the waters of the continental coasts are much shallower, owing to the gentle downward slope of the continental shelves. The waters above many continental shelves receive an influx of nutrients from several rivers, including St. Lawrence, Mississippi, and the Amazon in the Americas, the Rhine in Europe; and the Congo River in Africa. These nutrients support the growth of plankton and algae, the base of the marine food chain. Plankton, in turn, supports many species of mollusks, crustaceans, fish, and marine mammals that thrive in these waters.

Abyssal Plains and Deep Ocean Trenches

Beyond the continental shelves, the ocean floor consists of vast, flat expanses called abyssal plains. Many deep ocean trenches cut through the abyssal plains, especially in the Caribbean. The deepest point in the Atlantic Ocean is the Puerto Rico trench, some 28,000 feet deep, located at the junction of the North American and Caribbean plates. Less is known about marine life in these regions of the Atlantic. Organisms here feed on the remains of dead fish and whales sinking from the ocean's surface. Around hydrothermal vents, tube worms and other deep sea creatures depend on chemosynthetic bacteria as the base of the food chain since sunlight cannot penetrate these depths.

Mid-Atlantic Ridge

The Mid-Atlantic ridge is a chain of undersea peaks and volcanoes stretching over 10,000 miles from Iceland nearly to Antarctica. This ridge marks the fault line where the North and South American plates border the European and African plates. The seafloor here is spreading at the rate of 1 to 4 inches per year. In Iceland, the Mid-Atlantic ridge is visible above land. As the North American and European plates spread apart, magma from the Earth's mantle rises to the surface, accounting for Iceland's intense volcanic activity.

In the 1960s, scientists collected rock samples from both sides of the Mid-Atlantic ridge. Based on their identical structure and content of radioactive elements, researchers concluded that both sets of rocks were the same age, strongly suggesting a common origin. This discovery bolstered the theory of plate tectonics, an idea that fundamentally transformed the field of geology.

Gulf Stream

One of the major currents of the Atlantic Ocean is the Gulf Stream, a warm water current running from the Gulf of Mexico northward to the British Isles. The Gulf Stream carries tremendous volumes of warm, salty water (over 30 million cubic meters per second) from the tropics towards the Arctic Ocean. This water cools and sinks at high northern latitudes as it mixes with freshwater from arctic glaciers and the polar ice cap.

A countercurrent carries the cold water south toward the equator, where the cycle repeats. The heat given off by this so-called thermohaline conveyor is responsible for the mild climate of much of the North Atlantic.

For example, although the British Isles are located as far north as Hudson Bay, the climate is much warmer for most of the year.

The Sargasso Sea

The Sargasso Sea is a unique marine ecosystem located in the middle of the subtropical North Atlantic. Certain species of seaweed (called Sargassum) and leafy sea dragons are found nowhere else on Earth. The Sargasso Sea is a critical spawning ground for many species of eels and sea turtles, jackfish, dolphinfish, and the white marlin. The migration routes of tuna, loggerhead sea turtles, and humpback whales also cross the Sargasso Sea.

Essay 27: Geological Processes in the Earth's Interior

The Earth is a terrestrial planet consisting of a crust, mantle, molten outer core, and solid inner core. The Earth's crust is rich in silicon, aluminum, and other elements less massive than iron. Most of the Earth's iron and heavier elements are located in the mantle and core. The crust extends about 40 miles below the continents but only 10 miles underneath the ocean basins. Strange as it may seem, the crust comprises only 1% of our planet's mass; the mantle and core constitute the other 99%.

To put it in perspective, if the Earth were the size of a peach, the crust would correspond to the skin of the peach. The fruit and pit correspond to the mantle and core, respectively. The mantle consists of semisolid igneous rock that gradually becomes entirely liquid closer to the core. Small amounts of the mantle are ejected during volcanic eruptions, given that the roots of volcanoes often extend 100 miles or more below the Earth's surface. The outer core is thought to be made up of molten iron, nickel, and other metals. The inner core is an extremely dense metallic ball. Even though the inner core is hotter than the outer core, it remains solid due to its high density as well as the immense pressure exerted on it by the surrounding mantle and outer core.

All geological processes that occur below the Earth's surface are driven, directly or indirectly, by heat released from the Earth's core. According to geologists, most of the intense heat in the Earth's interior was produced by the accretion of the planet over 4 billion years ago. The remainder is produced by the continuous decay of radioactive elements.

As with other bodies, the Earth's heat is dissipated by radiation, conduction, and, most importantly, convection. Briefly, radiation means the emission of heat into surrounding space. This is the way Earth receives solar energy. Conduction means the transfer of heat from one object to another by direct contact. For example, a metal spoon placed in a cup of hot water will gradually become hot to the touch. Finally, convection refers to heat transfer by means of directly moving hot matter to a cooler region. The most well-understood example of geologic convection is volcanism, which is discussed next.

Volcanism: Hot Spots and Subduction Zones

Hot spot volcanoes result from the upwelling of magma plumes from the Earth's mantle, which penetrate through the crust and ultimately burst through the surface to form a volcano.

These hot spots are scattered across the continents and ocean basins. They include Yellowstone National Park (located atop a gigantic caldera volcano), the Galapagos, the Hawaiian Islands, and other volcanic peaks in Africa, the Arabian Peninsula, Indonesia, Siberia, Antarctica, and elsewhere.

Subduction zone volcanism occurs in regions where the ocean floor dives underneath an adjacent area of thick continental crust. As the oceanic crust melts into the underlying mantle, magma chambers are produced, resulting in chains of volcanoes. Examples of subduction zone volcanism include the Aleutian Islands off Alaska, Crater Lake, and Mt. St. Helens. Many volcanoes around the Pacific Ocean, the so-called Ring of Fire, were produced by subduction zone volcanism. In addition to volcanism, the Earth's mantle convection is also responsible for earthquakes and tsunamis.

Earthquakes

The Earth's lithosphere (crust and upper mantle) is fractured into 15 major tectonic plates, as shown on the diagram.

Heat circulating in the Earth's mantle is believed to drive the movement of tectonic plates. Although the plates are moving at approximately one inch per year, this accounts for the rearrangement of Earth's continents over vast spans of geologic time. At their boundaries, tectonic plates collide, spread apart, or move in parallel but opposite directions to each other. As a result of these movements, most earthquakes occur at tectonic plate boundaries. Seismic activity is a continuous phenomenon. However, earthquakes strong enough to be felt occur at intervals of months or years.

Tsunamis

Undersea earthquakes cause tsunamis. When large areas of the ocean floor suddenly collapse or are thrust up by tectonic plate movements, the result is a massive ripple effect below the ocean's surface. These waves may be undetectable in the open ocean without special monitoring equipment. As these waves approach islands and continental shelves, however, they gain amplitude quickly, with crests approaching 50 feet. When these waves crash onshore, they can devastate coastal regions, with death tolls in the hundreds of thousands, as happened in the December 2004 Indonesian tsunami.

Earth's Magnetic Field

Although magnetism is traditionally the domain of physics and astronomy, Earth's magnetic field is ultimately a product of geology. Many scientists think that the combination of a solid inner core surrounded by a molten outer core is necessary for the existence of Earth's magnetic field. Planetary bodies lacking this core structure, including Earth's moon, Venus, and Mars, have practically no magnetic fields. By studying the orientation of certain minerals in rock strata, geologists have determined that the direction of Earth's magnetic field reverses every 300,000 years or so. The basis of magnetic field reversal remains poorly understood.

Essay 28: Oceanic Heat Content

What causes changes in the ocean's heat content? This fundamental question represents an intense area of research in oceanography in particular and environmental science in general.

First, the earth's axis of rotation is tilted at a 23.5-degree angle. This tilt leads to seasonal variations in day length and, by extension, solar energy received on all parts of the planet except the equator. Second is the amount of freshwater frozen in glaciers and the polar ice caps. Ice has a profound effect on the heat content of the ocean. Intact sea ice reflects sunlight back into space. Astronomers call this phenomenon albedo. As sea ice dwindles (in all likelihood due to manmade global warming), less sunlight is reflected, more heat is absorbed, and a vicious melting cycle ensues. Almost all major glaciers around the world have melted to some extent over the last century.

A third related factor is the ocean's salt content or salinity. Differences in salinity are believed to drive the deep ocean currents responsible for thermohaline convection. The best-known example of thermohaline convection is the Gulf Stream in the Atlantic Ocean. This massive flow of warm salt water is responsible for the temperate climate of Northern Europe, especially the British Isles. If the thermohaline conveyor were to shut down, Great Britain could experience a climate as frigid as that of the Hudson Bay in Canada.

Other factors no doubt influence oceanic heat content. Volcanic eruptions on land decrease the amount of solar energy reaching the ocean's surface. In contrast, undersea volcanoes add heat to their surroundings when they erupt. Long-term trends in global volcanism are difficult, if not impossible, to predict with current technology. Hence, their contribution to changes in ocean temperature can only be studied retrospectively. In this regard, ice core data obtained from Greenland and Antarctica has proved to be extremely valuable. Changes in the ratios of oxygen isotopes trapped in each layer of ice correspond to oceanic heating and cooling cycles.

Over long spans of geologic time, tectonic plate movements also affect ocean temperature. Without a nearly continuous ring of land around the Arctic Ocean to trap cold water, ice caps would not form over the Earth's North Pole. The position of the continent of Antarctica atop the South Pole sets up a similar situation in the Southern Hemisphere. This shifting geological arrangement, combined with changes in the tilt of the Earth's axis and the shape of the Earth's orbit over 26,000 to 100,000 years (Milankovitch cycles), are believed to be the driving force behind past Ice Ages. Millions of years from now, continental drift will rearrange the positions of the Earth's major landmasses, and Ice Ages may become a relic of Earth's distant past.

Essay 29: Rock Formation in Geology

Rocks remember; gases forget. This is a favorite maxim of geologists and planetary scientists. Essentially, this means that the composition of a rock yields important clues about its origin. In geologic terms, a rock is an aggregate of one or more minerals. Minerals, in turn, are composed of one or more elements. All rocks fall into one of three categories: igneous, sedimentary, and metamorphic.

Igneous rocks are formed by volcanic activity. The term 'igneous' literally means "fire," an apt description of the high temperatures responsible for the formation of these rocks. Because the Earth began in a molten state over 4 billion years ago, all of its rocks originated as igneous rocks. Geologists subdivide igneous rocks into two groups: intrusive and extrusive.

Intrusive igneous rocks solidify below the Earth's surface, including granite, diorite, pegmatite, and peridotite. Extrusive igneous rocks solidify at or above the Earth's surface. They include rocks found in lava flows, such as basalt, obsidian, and pumice. Most of the rock found in the oceanic crust consists of basalt. Obsidian forms when lava cools extremely fast. It contains no crystals, and, like glass, its pieces have exquisitely sharp edges. Pumice is far less dense than obsidian and is commonly used as an abrasive.

Today, new igneous rocks are formed by the upwelling of magma from the Earth's mantle during volcanic eruptions as well as in ocean basins where the seafloor is spreading. Another source of igneous rocks comes from the melting of crustal rocks as they are pushed into the mantle during tectonic plate subduction.

Sedimentary rocks are formed when layers of sediment become cemented together over long periods of time. Common examples of sedimentary rocks include shale, flint, sandstone, and limestone. Sedimentary rocks often contain the fossils of marine organisms, which incorporated calcium carbonate into their shells during their lifetimes. When these organisms died, their shells sank to the ocean floor, where they were literally cemented into the surrounding sediments. Many sandstone and limestone deposits contain the fossilized remains of trilobites, ammonites, and other long-extinct sea creatures.

Metamorphic rocks are produced when igneous or sedimentary rocks are transformed by high temperatures and pressures or undergo chemical reactions. Approximately 27 percent of the Earth's crust is thought to consist of metamorphic rock. Since the Earth's crust comprises a mere one percent of the planet's total mass, however, the vast majority of Earth's rocks are either igneous or sedimentary.

An easy-to-remember example of a metamorphic rock is marble, found in caves (and monuments) throughout the world. Marble forms when porous limestone (composed largely of calcium carbonate) is heated and squeezed until it recrystallizes into a denser configuration. Under a microscope, limestone is packed together loosely, whereas marble fragments are locked together like jigsaw puzzle pieces. Other metamorphic rocks include amphibolite, gneiss, quartzite, schist, slate, and soapstone.

Geologists broadly categorize metamorphic rocks as foliated or non-foliated. Foliated metamorphic rocks exhibit distinct bands or layers because their crystals are formed from multi-mineral sedimentary rock exposed to directional pressure. A good example of a foliated rock is slate. In contrast, non-foliated rocks show little or no layering because their crystals were exposed to pressure from multiple directions. Examples include marble and quartzite.

Essay 30: Soil Quality – Coffee Grounds, Eggshells, and Earthworms

Author's disclaimer: This article contains advice based on common sense and anecdotal evidence. Individual results may vary, but as the saying goes, nothing ventured, nothing gained.

"Bad" soil may be of poor quality for a number of reasons. First, the soil may be depleted of potassium, nitrogen, and phosphate nutrients. Nutrient depletion tends to occur after several consecutive years of growing the same crop without utilizing proper fertilizer or crop rotation techniques. Second, the soil may have a high salinity level, as often occurs in arid areas subject to excessive irrigation. Third, the topsoil itself may have disappeared due to wind and/or water erosion.

In the latter two scenarios, no easy remedies exist. Removing the ruined soil and replacing it with fresh topsoil and mulch may be feasible for small plots of land such as backyard gardens. Still, anything larger than a few acres quickly becomes a cost-prohibitive project. As such, the remainder of this article will focus on the first scenario - nutrient depletion in salvageable soil.

Based on the experiences of relatives and close friends, one technique for enriching soil indefinitely is by making it hospitable for earthworms. These creatures directly enrich the soil by ingesting and excreting it and indirectly by returning carbon and nitrogen to the soil when decomposing. Earthworms also create air pockets in the soil as they move through it; this helps break up dry clods, allowing plants to extend their roots deeper into the soil.

How does one attract earthworms in the first place? The answer is by enriching the soil with organic waste, especially materials like coffee grounds, banana peels, and spoiled vegetables. These, along with animal manure, make excellent compost. After applying the compost mixture, earthworms may migrate on their own from surrounding areas. Otherwise, one can substitute night crawlers available at most bait and tackle shops.

In addition to the earthworm approach, another material that will enrich the soil is crushed eggshells. This is a guaranteed way to replenish calcium and phosphorus stores. For what it's worth, adding milk or urine to soil contributes nitrogen in the usable forms of protein or urea, respectively.

One final suggestion is to scatter small pieces of chalk or limestone on top of the soil. Over time, rainwater will leach minerals out of these rocks, slowly enriching the soil in the process. These techniques should do the trick; however, if all else fails, there is no shame in purchasing fertilizer and potting soil from a home and garden store.

Human Anatomy & Physiology

Essay 31: Amino Acid Metabolism Pathways

In the realm of biochemistry, 'metabolism' comprises two distinct categories: anabolism, or synthesis, and catabolism, or breakdown. In humans, twenty amino acids serve as the building blocks of all proteins. Amino acids consist of a central carbon bonded to an NH2 group (the amino terminus), a COO- group (the carboxyl terminus), and a variable ligand called a side chain.

Although amino acids can be categorized in a number of ways, most scientists focus on the properties of their side chains. The aliphatic amino acids include alanine, glycine, leucine, isoleucine, proline, and valine. They have relatively hydrophobic, chemically inert side chains. Three amino acids are commonly phosphorylated on their -OH groups: serine, threonine, and tyrosine (also an aromatic amino acid). Two amino acids contain sulfur cysteine and methionine. Two other amino acids contain non-reactive aromatic side chains - phenylalanine and tryptophan. The acidic amino acids are glutamate and aspartate; their amine derivatives are glutamine and asparagine. Amino acids with basic side chains include arginine, histidine, and lysine.

Since amino acid catabolism in humans is arguably more complicated than anabolism, it will be discussed first. Amino acids present a special challenge for the human body. Unlike carbohydrates and lipids, they cannot be stored for later use. Amino acids must be used for protein synthesis, broken down for energy, or, in some cases, converted to other useful molecules such as neurotransmitters.

The definitive step in amino acid catabolism involves the removal of the nitrogen-containing amine group, a process called deamination. Since humans cannot dispose of nitrogen as ammonia due to its extreme toxicity, ammonia enters the urea cycle, where liver cells conjugate it into urea. The bloodstream carries urea to the kidneys, which excrete it in the urine.

The remainder of the amino acid is referred to as a carbon skeleton, which biochemists classify as either glucogenic or ketogenic.

Glucogenic amino acids undergo a series of enzymatic reactions, which convert them to pyruvate or intermediates of the Krebs cycle; these intermediate compounds can ultimately be turned into glucose by a process called gluconeogenesis. Eighteen of the twenty amino acids fall into this category.

Ketogenic amino acids give rise to a group of molecules called ketone bodies, specifically acetoacetate and beta-hydroxybutyrate. Ketone bodies are converted into a two-carbon unit called acetyl CoA, as opposed to pyruvate. Consequently, their carbon skeletons cannot be converted to glucose (because human cells cannot convert acetyl CoA back into pyruvate.) The only purely ketogenic amino acids are leucine and lysine. Five amino acids are both glucogenic and ketogenic: isoleucine, phenylalanine, tyrosine, tryptophan, and threonine. All other amino acids are purely glucogenic.

Minor metabolic fates of amino acids are diverse, and the most important ones are discussed.

Glycine is used as a precursor in the synthesis of heme and cholesterol. It also acts as an inhibitory neurotransmitter in the spinal cord.

Tyrosine is the source of thyroid hormones as well as the pigment melanin made by melanocytes. It is also the precursor of the catecholamine neurotransmitters dopamine, norepinephrine, and epinephrine, produced by neurons in the central nervous system (CNS) and by chromaffin cells in the adrenal medulla.

Tryptophan is converted to serotonin (5-hydroxytryptamine or 5-HT) in the intestinal tract's CNS neurons and neuroendocrine cells. Serotonin affects gut motility, sleep, and mood.

The pineal gland converts serotonin to melatonin, which helps regulate the sleep-wake cycle and allows the brain's circadian rhythms to adjust to changes in day length.

Histidine is decarboxylated to form histamine. This reaction occurs in the CNS as well as in specialized intestinal cells called ECL cells. Basophils and mast cells store histamine inside cytosolic granules and release it during allergic reactions.

Lysine is the precursor of carnitine, a carrier molecule required for the transport of long-chain fatty acids into the mitochondrial matrix.

Glutamate is the main excitatory neurotransmitter in the CNS. Some neurons decarboxylate glutamate to form GABA (gamma amino butyric acid), the nervous system's major inhibitory neurotransmitter. Aspartate also serves as an excitatory neurotransmitter, mainly in spinal cord neurons.

Methionine is incorporated into a carrier molecule called SAM (S-adenosylmethionine), which plays a key role in the transport of methyl groups.

Genetic diseases involving amino acid catabolism (sometimes called inborn errors of metabolism) are relatively rare but can have devastating consequences. Phenylketonuria (PKU), caused by a defect in phenylalanine breakdown, occurs most often in Celtic populations, at a frequency of 1 in 25,000 newborns in the U.S. Infants with PKU must be placed on a special diet in the first six months of life, or severe mental retardation results.

Less common but far more severe is maple syrup urine disease, which results from defective metabolism of the branched-chain amino acids leucine, isoleucine, and valine. This disorder is seen almost exclusively in Amish populations, in an estimated 1 out of 380 births, as opposed to 1 in 185,000 births in the general population. Children with maple syrup urine disease seldom survive to adulthood, even with treatment.

An inability to metabolize glycine results in non-ketotic hyperglycinemia, or NKH, a disorder that occurs most often in people of Finnish ancestry (1 in 55,000 births) and is marked by seizures starting in infancy. Finally, a defect or absence of any urea cycle enzyme results in hyperammonemia, a condition characterized by seizures and, if untreated, coma and death. Urea cycle defects occur relatively rarely in an estimated 1 out of 30,000 newborns.

Anabolism of amino acids is far less complicated in humans compared to many other organisms. Most plants, fungi, protozoa, and virtually all bacteria can synthesize all 20 amino acids from simple carbon and nitrogen-containing precursors. Adult humans, on the other hand, must obtain 9 amino acids entirely from the diet.

These are known as essential amino acids, and a deficiency of even one of them can result in protein malnutrition, also called a negative nitrogen balance.

The essential amino acids include phenylalanine, valine, tryptophan, threonine, isoleucine, methionine, histidine, leucine, and lysine. Arginine is an essential amino acid in children, but adults can obtain a sufficient amount from the urea cycle to avoid a negative nitrogen balance. A popular mnemonic for remembering the essential amino acids is PVT TIM H(A)LL.

The non-essential amino acids are synthesized in the liver, mainly from the metabolic intermediates of glycolysis, the Krebs cycle, and the urea cycle. Pyruvate and oxaloacetate are the precursors of alanine and aspartate, which in turn gives rise to asparagine.

Another molecule produced in glycolysis, 3-PG, gives rise to serine, cysteine, and glycine. Glutamate is derived from the transamination of the Krebs cycle molecule alpha-ketoglutarate; glutamate can be converted to glutamine and proline.

As previously mentioned, arginine is produced in the urea cycle, where it acts as the immediate precursor of urea. Finally, tyrosine can be formed from the hydroxylation of phenylalanine. Technically, this makes tyrosine a non-essential amino acid; however, this holds true only if a person's diet contains enough phenylalanine to divert some for tyrosine production.

Essay 32: The Amygdala

The term amygdala refers to a bilateral cluster of nuclei located deep within the brain's temporal lobes. The word amygdala itself is derived from the Greek word for almond. Apparently, the shape of these neuronal clusters vaguely resembles an almond, at least in the eyes of neuroanatomists.

The amygdala is usually considered to be part of the limbic system, a region of the brain involved with emotion, autonomic reactions, and memory. In addition to the amygdala, the limbic system includes the olfactory tracts, cingulate gyrus, hippocampus, anterior thalamus, and hypothalamus. The amygdala is thought to regulate the emotional content of memory in humans as well as other mammals.

The Papez circuit

Over 70 years ago, a scientist named George Papez proposed that the limbic system structures act as a circuit to integrate human emotional and physiological reactions to various stimuli. The output of the limbic system manifests as stereotyped, reflexive behaviors that become hardwired into long-term memory.

For example, many people have a strong fear of snakes and spiders. Regardless of the reason for this phenomenon, the sight of a snake or spider (or even viewing a photograph of one) triggers a predictable set of responses. The person's heart rate rises, as does blood pressure; the pupils and airways dilate; digestion stops; and the person may start to sweat or even hyperventilate. Clearly, visual stimuli can evoke a fight/flight response - the domain of the sympathetic nervous system - as well as the emotions that accompany this response, such as fear, anger, and a heightened sense of alertness.

Animal models

Few, if any, people have ever suffered brain damage exclusively affecting the amygdala. As such, scientists have turned to animal models to understand the amygdala's role in behavior better. Predictably, scientists were curious to find out what would happen in animals following ablation (surgical destruction) of the amygdala. In mice, the result was a complete loss of their normal fear response. Whereas the sight of a cat would cause normal mice to flee, mice lacking a functional amygdala approached cats and other predators nonchalantly.

Chimpanzees subjected to bilateral destruction of the amygdala developed a complex pattern of abnormal behaviors called Kluver-Bucy syndrome after the researchers who first reported these findings. Similar to the mice, the chimpanzees reacted in bizarre ways to their surroundings. For example, when a hungry chimpanzee is offered food along with inedible items, it immediately eats the food and ignores the other objects. Instead, the Kluver-Bucy chimpanzees leisurely examined each object, visually and orally, as though they had lost all sense of object recognition, a condition called psychic blindness. The chimpanzees also exhibited bouts of hypersexual behavior and unprovoked aggression.

In humans, the closest parallel to Kluver-Bucy syndrome is damage to the temporal lobes secondary to trauma, uncontrolled seizures, viral encephalitis, or a brain tumor. Patients with these conditions may exhibit blunted emotions or, alternatively, display a labile affect, manifesting as paranoia and a hair-trigger temper. People with these aberrant emotional responses sometimes benefit from treatment with mood stabilizers or neuroleptic medications.

Essay 33 The Biological Basis of Memory

How does the brain form and retain memories? This question has fascinated people from all walks of life from time immemorial. Although scientists have discovered a tremendous amount about brain function over the past century, the biological basis of memory largely remains a mystery. Part of the problem, however, is the nature of the question itself. This leads to a more basic question: what does the term memory really mean?

In the scientific realm, the term 'memory' encompasses far more than a person's ability to recall facts or past events. This ability is called declarative memory. The main subdivisions of declarative memory are short-term and long-term memory. Anyone who has ever pulled an all-nighter cramming for a final exam is well aware of this distinction. In contrast, remembering how to perform a series of motor skills is known as procedural memory. Spatial memory is rather tricky; it combines some aspects of declarative memory with a mental map of the external environment. The ability to keep track of long-term goals is the hallmark of executive memory. This article will touch on each aspect of memory and hopefully shed some light on this fascinating phenomenon.

In some respects, short-term memory is the most straightforward aspect of declarative memory to study. The vast majority of information the brain processes probably makes it no farther than short-term memory. For example, if you see an unfamiliar phone number on TV, you can usually remember it long enough to jot it down. However, you will most likely forget it if a few minutes elapse without you seeing the number or mentally repeating it. Another good example is trying to remember what you ate for dinner last night as opposed to last week. Unless last week's dinner was a special occasion (or you eat the same meal every night), you probably will not remember the event's details; however, you probably will have a vague recollection of the event itself.

Clearly, some information makes it through the filter of short-term memory and ends up in the vast reservoir of long-term memories. In 1949, a Canadian neuroscientist named Donald Hebb proposed that the key to understanding memory was a process called long-term potentiation (LTP). This concept is often expressed as the maxim "neurons that fire together wire together."

By this, he meant that repeated stimulation of two neurons reinforces their synaptic connections, resulting in stronger responses to subsequent stimuli. Conversely, a lack of stimulation leads to weakened synaptic connections and sometimes to their complete disappearance.

At the time, the idea that neuronal synapses exhibited plasticity throughout adult life was a fairly radical concept. Some scientists were skeptical of the notion that brain circuits were malleable after infancy. Others were intrigued by the concept but were unsure where in the brain to look. A few years later, an important clue would appear in the form of Henry Molaison, known by his initials H.M., whose case definitively established the role of the brain's hippocampus in consolidating short-term information into long-term memory.

The Case of H.M.

One of the most famous patients in the annals of medicine, H.M. (1926-2008) was a dramatic example of the consequences of bilateral destruction of the hippocampus. After sustaining a head injury in childhood, H.M. developed temporal lobe epilepsy, which grew worse over time. When no medications helped, he underwent radical brain surgery as a last ditch effort to control the seizures. Employing a technique called electrical ablation, surgeons destroyed most of H.M.'s deep temporal lobes, and with them his hippocampi.

For readers interested in etymology, the hippocampus earned its name due its vague resemblance to a seahorse. Hippocampal neurons are connected to several parts of the brain involved with emotions and arousal including the hypothalamus and amygdala. A series of axons called Schaffer collaterals forms a strong excitatory circuit within the hippocampus itself. Although this part of the brain readily undergoes LTP, the possibility of an out of control excitation loop means the hippocampus is also the most likely place in the CNS for a seizure to originate.

The operation successfully ended H.M.'s seizures; unfortunately, it also ended his ability to form long term memories. H.M. developed severe anterograde amnesia immediately after the operation, to the extent that he was unable to remember any new information for longer than a few seconds. When asked his age, H.M. would consistently underestimate it. When asked his place of residence, H.M. could only remember his childhood address. Interestingly, H.M.'s procedural memory remained intact. For instance, he became adept at tracing figures; on the other hand, he never recalled practicing this activity let alone having been asked to learn this skill in the first place.

In contrast to short term memory, surprisingly little is known about long term declarative memory. The main points of agreement are, first, the brain somehow sifts through the vast stream of conscious experience, retaining information that seems interesting, important, or both. Second, information encountered repeatedly can be over learned to the point that forgetting it becomes virtually impossible.

For example, nearly all Americans can identify the figure on the front of the one dollar bill as George Washington. When asked about Washington's significance, the invariable response is that he was the first President of the United States. (As an aside, almost no one distinctly remembers the first time s/he learned these two facts). Even people with severe dementia tend to remember the name of the first U.S. president. A general consensus has emerged that long term memories (or over learned facts at least) seem to be stored in multiple areas of the cerebral cortex. In light of these observations, no alternative explanation seems plausible.

Procedural or motor memory seems to be the domain of the cerebellum. Specifically, a process called LTD (long term depression) in the Purkinje cell layer of the cerebellar cortex seems to fine tune motor programs. In practical terms, this means that when it comes to riding a bicycle or playing video games, individual skills improve to a point then level off. Cerebellar damage often results in loss of fine motor control, disequilibrium, as well as apraxias, or difficulty completing a planned series of movements.

Spatial memory

The most widely used method to assess an animal's spatial memory is a test called the Morris water maze, which consists of a small container filled with cloudy water and a submerged platform on which a mouse can fit. Normal mice figure out the platform's location after a few minutes and remember its whereabouts during subsequent trials. Drugs that block protein synthesis, however, also block a mouse's ability to remember the platform's location.

In humans the brain's right parietal lobe is especially important in spatial memory. Some stroke victims experience a condition called parietal neglect syndrome. When given a circle and asked to draw a clock face, they crowd the numbers into one half of the circle. Some patients shave only half of their face. Apparently, in their brains' mental map, half of the world has ceased to exist.

Executive memory

In a sense, this is the most difficult form of memory to pin down. One example would be asking a student the type of degree s/he expects to receive upon graduating from a 4 year undergraduate program. Almost all would reflexively answer "a bachelor's degree," even though they might go days at a time without thinking about school, or, for that matter, attending class. The main insights into executive memory come from rare cases of patients with dissociative or psychogenic fugue. This term describes people who, after intense emotional stress or trauma, or sometimes for no apparent reason, abandon their jobs, friends, and family; in essence their identity evaporates. They are sometimes rediscovered weeks or months later in a distant location with a different identity and no recollection of the events leading to their disappearance.

Essay 34: How does the brain sense osmolality and regulate water balance?

Before answering this question, it is necessary to understand the meaning of the term osmolality as well as its implications for human physiology.

In chemistry, the osmolality of a solution is defined as the number of moles of solute (e.g., salt or sugar) per kilogram of solvent. When the solvent in question is water, osmolality is practically the same as the more familiar unit osmolarity, defined as moles of solute per liter of solvent. Since 1 liter of water has a mass of approximately 1 kilogram, the two values are virtually the same.

In the case of the human circulatory system, approximately 55% of blood volume is composed of straw-colored plasma, while the remainder is occupied by the so-called formed elements of blood: red blood cells, white blood cells, and platelets. Plasma itself is 90% water and contains many dissolved components: glucose, electrolytes (sodium, chloride, and bicarbonate ions), as well as a variety of proteins, including albumin, clotting factors, and soluble antibodies.

All of these plasma components contribute to the osmolality of blood, the most important being sodium and glucose. Physiologists have assigned a unit to blood osmolality called the milliosmole, abbreviated mOsm. As with all other metabolic parameters, the body must maintain serum osmolality within a safe range, from 280 to 303 mOsm per kilogram of body weight. In other words, the body cannot allow the blood to become either too dilute or too syrupy.

Three brain sensors regulate osmolality.

Certain regions of the brain, known as the circumventricular organs, lack a blood-brain barrier. In a sense, they act as the brain's eyes and ears, allowing it to assess metabolic parameters and make necessary adjustments quickly. In humans, two of these sensors are the subfornical organ (SFO) and the vascular organ of the lateral terminalis (OVLT), both located near the hypothalamus.

These clusters of neurons contain specialized surface proteins called osmoreceptors capable of sensing changes in the concentration of sodium and chloride ions. If the blood becomes too concentrated, the SFO and OVLT activate hypothalamic neurons, ultimately culminating in the sensation of thirst.

The third brain sensor is located in the hypothalamus and posterior pituitary gland. It consists of neurons whose cell bodies are located in the supraoptic (SON) and paraventricular nuclei (PVN) of the hypothalamus but whose axon terminals extend into the posterior pituitary. These cells also contain osmoreceptors, but to some extent, they rely on hormonal signals from the kidneys and adrenal glands to gauge osmolality.

Because the kidneys interpret low blood pressure as low blood volume, their default response is retaining sodium. This is accomplished by an elaborate system known as the renin-angiotensin-aldosterone axis. In addition to its renal effects, angiotensin II stimulates SON and PVN cells to release vasopressin, also called antidiuretic hormone (ADH), directly into the bloodstream.

ADH travels to the kidneys, where it acts on cells in the distal convoluted tubules and collecting ducts, causing them to insert special proteins called aquaporins in their membranes. Aquaporins are protein channels that selectively transport water out of the kidney tubules, where it finds its way back into the bloodstream, as opposed to being lost in the urine. Diabetes Insipidus (DI) and SIADH.

As you may have guessed, negative consequences ensue if the brain's osmolality sensors are damaged or become overactive.

Head trauma, tumors, or increased intracranial pressure can compress the posterior pituitary or destroy this structure altogether. Any of these situations lead to a disorder known as central diabetes insipidus (DI), in which ADH is not released into the bloodstream. The kidneys are unable to concentrate urine, and the person may lose over a gallon of fluid a day in the form of dilute urine.

Central DI usually responds to desmopressin, a synthetic form of ADH which may be administered intranasally or taken orally. In contrast, in nephrogenic DI, the kidneys themselves are unresponsive to ADH. Nephrogenic DI may respond to thiazide diuretics, although sometimes the only effective treatment is to increase fluid intake to match urine output.

Excessive release of ADH is termed SIADH or Syndrome of Inappropriate ADH release. This condition seldom occurs as an isolated event. For poorly understood reasons, SIADH may occur in the context of pneumonia as well as a paraneoplastic syndrome in certain malignancies, most notably small-cell lung cancer. Most cases of SIADH respond to diuretics or lithium, which induces urination as a side effect.

Essay 35: Cell Junctions

Cell junctions are present in humans and all multicellular organisms and serve two essential but opposite purposes. First, they allow materials to travel freely from one cell to another, as with gap junctions. In contrast, another set of structures called desmosomes, found in desmosomal and tight junctions, create a largely impermeable barrier between certain tissue types, for example, the blood-brain barrier as well as the epidermis and dermis of the skin. Gap junctions, desmosomes, and tight junctions are the most well-characterized cell junctions in humans (aside from cell-cell adhesion proteins), and this article will focus on these three in turn.

Gap junctions consist of proteins called connexons. These proteins form hexametric (six subunits) complexes which align on the plasma membranes of the two adjacent cells. When these channels open (often in response to an electrical discharge), any molecule large enough to fit through the central pore of the gap junction can move from one cell into the other. Aligned gap junction proteins form structures aptly called electrical synapses, which have two conformations: open, corresponding to the on state, or closed, corresponding to the off state. Electrical synapses mediate the tail twitch reflex seen in goldfish.

In humans, gap junctions are found mainly in cardiac muscle as well as parts of the central nervous system. This makes perfect sense from the standpoint of cardiac physiology. Because gap junctions connect cells in the heart, calcium can reach all parts of the ventricles within a few milliseconds. This virtually guarantees synchronous muscle contractions, even in the case of cell-to-cell excitation (as in bundle branch block). In the brain, glial cells, especially astrocytes, are connected by electrical synapses. Some CNS neurons may be connected by gap junctions as well.

Although the functions of gap junctions in the brain are less obvious than in the heart, neurobiologists speculate that they may contribute to the high speed of signal propagation in the CNS. Alternatively, the major role of CNS gap junctions may be to allow astrocytes to efficiently recycle neurotransmitters like glutamate while simultaneously maintaining extracellular ionic gradients. The debate continues.

In contrast to gap junctions, desmosomes anchor cells together closely enough to create a water-resistant barrier. The best-understood example of a desmosomal junction in humans is found in the epidermal layer of the skin. Here, desmosomes bind to keratin and other intermediate filament proteins, contributing to human skin's mechanical strength.

The so-called blood-brain barrier exemplifies the third type of cell junction, the tight junction. Tight junctions consist of a series of adherens proteins (consisting of desmosomes bound to each other), which essentially staple the two cells together along these seams. In humans, tight junctions are found between the small intestine's enterocytes and, most notably, at the blood-brain barrier.

The term blood-brain barrier arose when scientists noticed that fresh human brain tissue failed to take up most stains that other tissues absorbed quite easily.

With the advent of electron microscopy, the basis of the blood-brain barrier was elucidated. Capillary endothelial cells in the brain and spinal cord are connected by tight junctions and surrounded by the podocytes (foot processes) of glial cells called astrocytes. By contrast, the capillary beds in most body tissues are relatively leaky. Unlike the liver or kidneys, however, the brain lacks lymphatic vessels for fluid drainage and is off-limits to most cells of the immune system.

As such, it makes sense to confine water, microbes, and most other chemicals to the blood circulation. Nonetheless, the brain needs a steady supply of glucose and amino acids. To circumvent the blood-brain barrier, cells in the brain contain specialized membrane channels to import acidic, neutral, and basic amino acids. CNS cells also contain glucose transport proteins that work independently of the hormone insulin.

The blood-brain barrier is permeable to several small molecules, most notably ethanol and ammonia. Lipophilic compounds such as organophosphate insecticides and nitrosoureas (mustard gas) also cross the BBB, as do toxic doses of heavy metals such as lead, mercury, arsenic, and platinum - sometimes in the form of chemotherapy drugs like arsenic trioxide and cisplatin.

Essay 36: Cholesterol Explained

At this point in time, the word 'cholesterol' has a decidedly negative connotation, conjuring up images of gallstones, heart attacks, and strokes. Although excess cholesterol often plays a role in these disease states, that is only part of the story. This article will focus on what cholesterol is and why the human body needs a certain amount of it to function properly.

Cholesterol is a small, hydrophobic molecule composed almost entirely of carbon and hydrogen atoms. The 27 carbon atoms in cholesterol are all derived from the molecule acetyl CoA (coenzyme A), synthesized from the breakdown of amino acids, sugars, and fats. Acetyl CoA is often channeled into the Krebs cycle in order to harvest its chemical energy; however, it is also used as the initial compound in many biosynthetic pathways.

In the case of cholesterol biosynthesis, acetyl CoA is first converted to a compound called hydroxymethylglutarate CoA or HMG CoA. At this point, an enzyme called HMG CoA reductase catalyzes the committed step in the pathway, turning HMG CoA into mevalonic acid. From there, a series of enzymes convert mevalonic acid into squalene, a linear intermediate that folds to yield cholesterol's four-ringed structure.

Almost all cholesterol synthesis occurs in the liver, mainly because the liver is also the site of bile acid production. The liver produces enough cholesterol in adults to meet the body's needs. Dietary cholesterol comes in the form of dairy products, eggs, red meat, and shellfish.

Cholesterol has three main functions in the human body:

1) First and foremost, cholesterol stabilizes cell membranes by decreasing the fluidity of the lipid bilayer. Without cholesterol interspersed in the cell membrane, human cells would literally come apart at the seams.

2) Bile production - cholesterol is the precursor of bile acids produced by the liver, including cholic, chenodeoxycholic, and deoxycholic acid. Some of these acids are conjugated with glycine and taurine to increase their water solubility. Bile acids are then mixed with the breakdown products of dead red blood cells (mainly bilirubin) to produce a greenish-brown liquid called bile. Once bile is released from the hepatocytes, it travels down the biliary ducts toward the duodenum of the small intestine. Excess bile flows into the cystic duct and ultimately into the gallbladder, where it is stored between meals.

Bile emulsifies fats, breaking them into smaller globules, thereby exposing a greater surface area to the lipase enzymes responsible for fat digestion.

3) Steroid hormone and vitamin D synthesis - Cholesterol serves as the precursor of all steroid hormones as well as vitamin D, which is necessary for dietary calcium absorption. Steroid hormones include cortisol, a glucocorticoid; aldosterone, a mineralocorticoid; estrogen, and testosterone. Aside from sex hormones produced mostly in the gonads, the other steroid hormones are synthesized by cells in the adrenal cortex.

Vitamin D acts in a manner analogous to a steroid hormone. It is formed by the action of UV light on a cholesterol derivative found in the skin. This vitamin D precursor travels to the liver and ultimately to the kidney, which converts vitamin D to calcitriol, its most active form.

Calcitriol enters intestinal cells, acting as a transcription factor to induce the production of calbindin, a protein that traps dietary calcium. Dietary sources of vitamin D are relatively limited; they include fortified dairy products and cod liver oil.

Good cholesterol vs. bad cholesterol

No discussion of cholesterol would be complete without clarifying this misunderstanding. Due to the hydrophobic nature of cholesterol, it cannot travel through the bloodstream as a soluble molecule. As with other lipids, cholesterol must be transported by carrier proteins called apolipoproteins. These carrier proteins differ with respect to their overall size, cholesterol content, and how tightly they bind cholesterol. High-density lipoproteins (HDL) are believed to remove cholesterol from circulation and send it back to the liver for processing into bile acid or disposal from the body via the small intestine. For this reason, the popular press has dubbed HDL "good cholesterol."

In contrast, LDL (low-density lipoprotein) is thought to have the opposite effect, depositing cholesterol along the walls of arteries and initiating the formation of atherosclerotic plaques. Predictably, LDL has been maligned as "bad cholesterol," although the cholesterol it carries is identical to that carried by its benevolent counterpart, HDL. In addition to HDL and LDL, other lipoproteins transport cholesterol, including chylomicron remnants, IDL (Intermediate Density Lipoprotein), and VLDL (Very Low-Density Lipoprotein), but their role in the pathogenesis of vascular disease remains unclear.

Essay 37: Diet and Metabolism

The link between a balanced diet and metabolism is paradoxically less direct and, at the same time, more complicated than most people imagine. On the one hand, consuming sufficient calories to meet the body's energy needs, in addition to essential amino acids, fatty acids, vitamins, and minerals, is enough to sustain human metabolism. The body can even store enough calories in its fat reserves to survive 2 to 3 months of total starvation. On the other hand, many endocrine disorders, especially of the thyroid gland, can severely disrupt metabolism even in the context of a balanced diet. This essay will examine the perceived connection between diet and metabolism in health and disease states.

It is common knowledge that growing children have a higher basal metabolic rate compared to adults. In this context, the term 'basal' means resting or baseline. Kilogram for kilogram, children require more protein and more total calories than adults simply because tissues undergoing active growth require more nutrients than tissues that are being maintained. A good analogy would be the energy requirements of someone swimming laps in a pool versus someone treading water. Almost everyone's metabolic rate declines after puberty, although people with low body fat and/or a high muscle mass tend to have a higher resting metabolism than their obese counterparts.

Disorders of the thyroid gland, adrenal glands, and occasionally the pituitary illustrate the indirect nature of the link between diet and metabolism. The metabolic alterations in diabetes are complicated enough to merit their own article. In hyperthyroidism, a person who loses weight despite maintaining the same diet often exhibits heat intolerance; if the condition goes untreated, he/she develops bulging eyeballs (proptosis) and cardiac arrhythmias.

The leading cause of hyperthyroidism is an autoimmune disorder called Grave's disease. In this case, autoantibodies bind to and activate TSH receptors on the thyroid follicular cells whether or not thyroid hormone is present. This effectively pushes the thyroid into overdrive, resulting in the above-mentioned symptoms. Occasionally, a pituitary adenoma releasing excess TSH (thyroid stimulating hormone) is the cause of hyperthyroidism. Even less commonly, an ovarian tumor called struma ovarii can produce enough ectopic thyroid hormone to induce hyperthyroidism.

The opposite scenario ensues in hypothyroidism. Patients exhibit weight gain, cold intolerance, and a slow heart rate. This condition usually results from a different autoimmune disorder called Hashimoto's disease, although an insufficient dietary intake of iodine causes some cases. Since thyroid hormone production requires iodine, people living far from the sea, e.g., in the Himalayan mountains, are susceptible to iodine deficiency. Iodized salt is available in most areas of the world today, making iodine deficiency a relatively rare condition.

Cushing's syndrome is the main adrenal disorder that disrupts metabolism, marked by truncal obesity, hypertension, diabetes, easy bruising, and osteoporosis. This condition is most often the result of long-term treatment with exogenous glucocorticoids, especially in people with severe asthma or lupus. A cortisol-secreting adrenal tumor may also cause Cushing's syndrome. In rare cases, the precipitating cause may be a pituitary adenoma secreting high levels of ACTH, which in turn drives excess cortisol release from the adrenal cortex. Strictly speaking, if the culprit is excess ACTH, the condition is referred to as Cushing's disease, even though the clinical manifestations may be identical. People with small cell lung cancer sometimes develop Cushing's disease in response to ectopic ACTH released by this tumor.

All in all, diet and metabolism share a connection but often lay worlds apart.

Essay 38: The Human Eye

The human eye is often compared to a video recorder or camera. Although certain similarities exist, the eye is far more complex than either of these machines. Part of the reason is that the main job of the eye is to provide the brain with the vast majority of its sensory input. Compared to other sensory modalities, the importance of vision is difficult to overemphasize. Scientists think that over 40% of the neurons in the brain's cerebral cortex are involved in analyzing visual information.

Much of the anatomy of the human eye is designed to nourish and protect the retina. The retina's photoreceptor cells are the only cells in the human body capable of converting photons of light into neuronal signals; as such, the retina represents the very heart of the visual system. The retina is so important, in fact, that scientists consider it to be a part of the central nervous system.

A protective membrane covers the anterior surface of the eyeball called the conjunctiva, which also coats the inner lining of the eyelids. The tough, fibrous wall of the eyeball itself is called the sclera. Extraocular muscles attach to the sclera, making coordinated movements of the eyeballs possible. The cornea is continuous with the sclera and mainly serves to protect the eye from mechanical trauma. Corneal scarring or abnormal proliferation of blood vessels (neovascularization) sometimes necessitates a corneal transplant. Transplanted corneas are generally obtained from cadavers.

The eye's anterior chamber consists of the cornea, iris, ciliary body, muscle, and lens. The iris consists of pigmented connective tissue lying in front of the lens. Contraction or relaxation of the ciliary muscle adjusts the size of the pupil, which is nothing more than a clear space within the iris that regulates the amount of light entering the eye. The lens consists of stacks of crystallin proteins that project incoming light rays on the surface of the retina. The ciliary body produces a fluid called aqueous humor, which circulates throughout the anterior chamber. Most of the aqueous humor drains through an opening called the canal of Schlemm, eventually returning to the venous circulation.

The major disease processes affecting the eye's anterior chamber are glaucoma and cataracts. Glaucoma results from partial or total obstruction of aqueous humor drainage through the canal of Schlemm. Fluid backup leads to elevated intraocular pressure. Over time, the increased intraocular pressure damages the optic nerve, resulting in irreversible vision loss.

Because glaucoma tends to be an insidious, asymptomatic disease, there is virtually nothing an ophthalmologist can do to reverse this disease once vision loss has occurred. Intraocular pressure can be measured at regular intervals, however, and all adults, especially people with a family history of glaucoma, are urged to schedule an annual eye examination.

Cataracts tend to occur in later life. Approximately half of all Americans will develop a cataract by age 80. The main cause of cataracts seems to be a buildup of calcium and protein deposits on the lens surface, clouding it and resulting in blindness if left untreated.

Certain rare diseases, most notably galactosemia, can cause cataracts in newborns. A few drugs, such as quetiapine (Seroquel), raise a person's risk of developing cataracts. People who take these drugs long-term are advised to be examined by an ophthalmologist at six-month intervals.

The eye's posterior chamber consists of the vitreous humor, choroid, and retina. The vitreous humor is colorless and has the consistency of a thick gel. The vitreous humor is thought to lend mechanical support to the posterior chamber of the eye and help maintain the shape of the retina.

The choroid is a pigmented layer of epithelial cells in which the photoreceptors of the retina are anchored. The photoreceptors themselves cannot be replaced by mitosis, but as long as these cells remain viable, their parts can be replaced throughout a person's life.

The retina itself consists of multiple cell types organized into three main layers: 1) Photoreceptor cells: Rods are responsible for black-and-white vision in dim light (scotopic vision). Cones come in three varieties and mediate bright light/color vision (photopic vision). 2) Bipolar cells - these cells transmit impulses from the rods, cones, and horizontal cells to the retinal ganglion cells. 3) Ganglion cells - these cells receive input from bipolar cells as well as amacrine cells, basically auxiliary cells that help integrate and fine-tune the signal.

The axons of the ganglion cells form the optic nerve, which wires the eyeball to the brain. The optic nerves intersect in a structure called the optic chiasm. Most fibers cross over to the contralateral or opposite side of the brain from the eye of origin. 95% of each optic nerve's fibers innervate a group of thalamic neurons called the LGN (lateral geniculate nucleus), which in turn sends signals to the primary visual cortex in the brain's occipital lobe. Most of the remaining fibers innervate the suprachiasmatic nucleus (SCN), a small cluster of cells in the hypothalamus that regulate the body's circadian rhythms.

Several disease processes can affect the posterior chamber of the eye, almost all of which involve the retina. These include chronic hypertension, diabetic retinopathy, retinal detachment, macular degeneration, autoimmune diseases, especially lupus and multiple sclerosis (MS), uncommon cancers such as retinoblastoma and ocular melanoma, as well as various mitochondrial disorders.

Long-standing high blood pressure and diabetes wreak havoc on the small blood vessels that nourish the retina. Over time, the retina's blood supply is compromised, starving the cells of oxygen and nutrients and eventually killing them.

In people with diabetes, fragile blood vessels form in an effort to compensate for the inadequate blood supply. This condition is called proliferative retinopathy; unfortunately, these new vessels tend to leak and cause further damage to the beleaguered retina. Diabetic retinopathy, retinal detachments, and wet macular degeneration can be treated with laser therapy (photocoagulation) as well as injections of an antibody called bevacizumab (Avastin), which inhibits angiogenesis.

Flare-ups of autoimmune diseases are usually managed with corticosteroids. Cancers arising from or involving the eye necessitate enucleation of the affected eyeball. No specific treatment exists for degenerative retinal diseases such as dry macular degeneration, retinitis pigmentosa, or mitochondrial disorders, e.g., Leber's Hereditary Optic Neuropathy. In the future, stem cell transplants or photosensitive electrode implants may partially restore vision in people with these conditions.

Essay 39: Fat – So Much More Than an F Word ;-)

Lipids (commonly known as fat) serve three main functions in the human body: first, as an energetically cheap way to store metabolic fuel for long periods of time; second, as the main structural component of cell membranes; and third, as the precursor of steroid hormones and other molecules involved in cell signaling. Each of these roles will be discussed in turn.

Humans and other animals store some calories in the form of glycogen (starch) and proteins; however, the most important reservoir for long-term energy storage is fat, specifically triacylglycerol (also known as triglycerides). These molecules are made of three fatty acids linked by an ester bond to a three-carbon unit called glycerol. Triacylglycerols are stored as droplets in specialized fat cells called adipocytes. The average person stores enough calories in their adipocytes to survive two to three months of total starvation.

As a survival strategy, starch stored in the liver is consumed within the first 24 hours of starvation. If no food is ingested by then, elevated cortisol levels and other stress hormones mobilize the body's fat reserves. In addition, the liver begins to produce high levels of substances called ketone bodies, which the brain can consume for energy as an alternative to glucose. When fat reserves are exhausted, skeletal muscle protein is the last fuel source broken down for energy. A professor of mine once compared this to burning furniture in a last-ditch effort to avoid freezing to death.

Besides serving as an energy depot, fats also play a structural role in the body by forming the lipid bilayer component of cell and organelle membranes. In contrast to the triacylglycerol molecules inside adipocytes, lipids in the cell membrane contain a so-called head region with hydrophilic phosphate groups and two fatty acid tails.

The most common lipid found in human cell membranes is phosphatidylcholine. Other structural membrane lipids include phosphatidylserine, phosphatidylethanolamine, cardiolipin, and phosphatidylinositol (more on that shortly). Another important lipid, cholesterol, is found in the membranes of animal cells, decreasing the lipid bilayer's fluidity. Cholesterol also serves as the precursor of bile acids, steroid hormones, and vitamin D.

The third major role that fats play in the body is cell signaling. This can be broken down into three subcategories: a) intracellular signaling molecules derived from membrane lipids, chief among them inositol triphosphate (IP3); b) arachidonic acid derivatives including prostaglandins, prostacyclins, thromboxanes, and leukotrienes; and c) steroid hormones and vitamin D, both derived from cholesterol.

IP3 is produced in response to the actions of certain neurotransmitters, e.g., acetylcholine and norepinephrine, on their target cells.

Briefly, in response to an extracellular signal, a cytosolic enzyme called phospholipase C cleaves IP3 from a special membrane lipid called phosphatidylinositol inositol. IP3 binds to a calcium channel on the endoplasmic reticulum (ER), resulting in the transient release of calcium ions from the ER lumen into the cytosol. Calcium ions, in turn, have multiple effects, like triggering muscle contraction, neurotransmitter release, and the activation of several types of enzymes.

Arachidonic acid, derived from a membrane lipid called diacylglycerol (DAG), is the precursor of a group of molecules called autacoids, including prostaglandins, prostacyclins, thromboxanes, and leukotrienes.

Autacoids are involved in pain, fever, inflammation, vascular tone, and platelet aggregation. NSAIDs (Non-Steroidal Anti-Inflammatory Drugs) like aspirin and ibuprofen treat pain and fever by blocking the production of certain inflammatory autacoids, in particular prostaglandins and thromboxanes.

Finally, all steroid hormones are derivatives of cholesterol. These include cortisol, the main glucocorticoid; aldosterone, the main mineralocorticoid; and last but not least, the so-called sex hormones, chief among them testosterone, estrogen, and progesterone. The adrenal glands produce all these steroid hormones; however, over 90% of sex hormones are produced in the testes/ovaries.

Vitamin D is also a cholesterol derivative, formed when a person's skin is exposed to UV light. After further activation in the liver, Vitamin D acts on intestinal enterocytes to promote calcium uptake as well as in the kidneys to promote calcium reabsorption. A dietary deficiency of Vitamin D combined with a lack of exposure to sunlight leads to a disease called rickets. Relatively few foods are rich in vitamin D, with the exceptions of cod liver oil and, in modern times, fortified milk and dairy products.

Essay 40: Glucose and Sodium Homeostasis

Two of the most important homeostatic mechanisms in the human body are the hormonal axes regulating blood glucose levels and serum sodium levels. In the case of blood glucose, the components of the system include the endocrine pancreas, the liver, and, to a lesser extent, skeletal muscle, adipose tissue, and the adrenal glands. Maintaining sodium balance involves an equally elaborate array of organs; the kidneys, adrenal cortex, liver, central nervous system, and even the heart play a role in sodium homeostasis.

Glucose regulation – contrary to popular thought, the body can tolerate elevated glucose levels – at least for short periods of time – far better than an episode of hypoglycemia. Physicians define hypoglycemia as a blood glucose level less than 70 mg/dL. Below this concentration, patients often become agitated, dizzy, and confused. Suppose blood glucose continues to drop, especially below 50 mg/dL. In that case, patients become markedly sweaty, nauseated, unresponsive, and prone to seizures unless they can ingest sugar orally (or receive it parenterally by IV).

The other extreme, hyperglycemia, is marked by glucose spilling into the urine at concentrations above 200 mg/dL. At 300 mg/dL, people often feel constantly thirsty and fatigued and may complain of blurry vision. Above 400 mg/dL, a person's serum exhibits hyperosmolarity; in other words, it behaves increasingly more like syrup than like blood plasma. The long-term complications of hyperglycemia are well understood – these are the hallmarks of diabetes mellitus. Microvascular complications of DM include retinopathy, peripheral neuropathy, vulnerability to infections, poor wound healing, and progressive renal failure. Macrovascular complications include peripheral arterial disease (culminating in limb gangrene and amputations), myocardial infarctions (heart attacks), and strokes.

With the above in mind, how does the body maintain a blood glucose level between 70-120 mg/dL (the normal range)?

The place to begin is the Islets of Langerhans, which make up the endocrine pancreas. The islets contain two main cell populations – α (alpha) cells, which produce glucagon, and β (beta) cells, which produce insulin. β cells contain a special type of glucose transporter on their surface called GLUT-2 that acts as a glucose sensor. When glucose levels rise after a meal, the β cells detect this spike in glucose concentration. The cells respond by closing potassium channels and opening calcium channels. The net effect is to cause their cell membranes to depolarize in much the same way as a neuron about to fire an action potential. When calcium influx depolarizes the membrane above a voltage threshold (-30 mV), the cell releases granules containing insulin directly into the bloodstream – much like a neuron releasing neurotransmitters into the synaptic cleft.

Insulin circulates in its active form for 15-30 minutes. During this time, it acts on several different organs to promote glucose uptake from the bloodstream. In the liver, insulin promotes the conversion of glucose to its storage form - a starch polymer called glycogen. The liver can store a 24-hour supply of glucose as glycogen. Skeletal muscle also takes up glucose when insulin binds. In this case, the binding of insulin causes the insulin receptor to dimerize and activate a tyrosine kinase signaling cascade, culminating in the phosphorylation of a protein called Insulin Receptor Substrate. IRS causes GLUT-4 transporters to translocate to the skeletal muscle membrane. The GLUT-4 transporters move glucose into the cytosol, which can be used immediately for ATP production or stored as glycogen. Insulin acts on adipose tissue to promote the conversion of glucose into fatty acids for long-term energy storage as triacylglycerol.

In contrast to insulin, the body has multiple mechanisms for raising blood glucose, especially in response to stress.

During brief emergencies, chromaffin cells in the adrenal medulla and adrenergic neurons release the catecholamine hormones epinephrine and NE. These hormones trigger the so-called fight/flight response. Glucagon raises blood glucose levels between meals by activating liver enzymes, which convert glycogen back into glucose. The liver releases free glucose into the bloodstream as opposed to using it for fuel; the reason is that the liver preferentially uses fatty acids and ketone bodies to drive ATP production rather than glucose. Cortisol, a glucocorticoid hormone produced by the zona fasciculata of the adrenal cortex, mobilizes glycogen and fat reserves for hours or even days during an episode of stress.

The diagram below illustrates the actions of insulin and glucagon in glucose homeostasis.

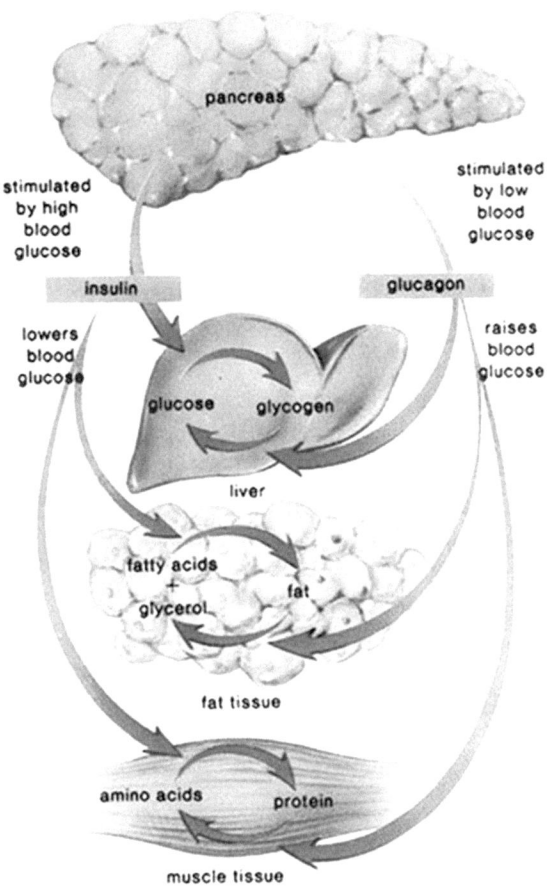

Image courtesy of: http://www.lifeextension.com/magazine/2004/6/report_diabetes/page-02?p=1

Sodium (Electrolyte) Balance

Physiologists often refer to sodium as the backbone of the extracellular fluid and emphasize that water follows sodium.

Although separate hormonal axes exist for water and sodium reuptake, sodium plays a crucial role in maintaining intravascular fluid volume in addition to contributing to serum osmolality. As will be explained, the kidneys conserve much of the water (a.k.a. plasma volume) in the glomerular filtrate by retaining sodium.

Serum sodium concentrations must be maintained between 130-145 mEq/mL. At levels below 130 mEq (hyponatremia), the CNS becomes swollen with water; cerebral edema leads to seizure, coma, and death if not corrected. At levels above 145 mEq (hypernatremia), nervous system function and cardiac and renal functions are disrupted. The body has developed an entire hormonal axis devoted exclusively to sodium homeostasis to avoid these extremes.

The kidney is the primary organ dedicated to sodium homeostasis; this means sodium retention in most scenarios.

Between the glomeruli and distal convoluted tubule (DCT) in each nephron, a sensor called the juxtaglomerular apparatus (JGA) measures the flow rate of sodium and chloride ions in the renal microcirculation as well as renal perfusion pressure. The former function is carried out by a cluster of cells called the macula densa, whereas the latter function is performed by granular cells acting as miniature baroreceptors. The JGA integrates these signals to estimate the body's extracellular fluid (ECF) volume. When the JGA detects a decline in ion count and/or perfusion pressure in the microvasculature, it interprets this as a loss of ECF volume. At this point, mesangial cells in the JGA release an enzyme called renin, the first step in activating the Renin/Angiotensin/Aldosterone system.

Renin cleaves a serum protein called angiotensinogen (produced in the liver) into Angiotensin I. As Angiotensin I flows through the renal and pulmonary circulations, a second enzyme called Angiotensin Converting Enzyme (ACE) cleaves Angiotensin I into Angiotensin II. Angiotensin II acts in three ways to conserve ECF volume. First, AT-II is a powerful vasoconstrictor. AT-II constricts the renal arteries and arterioles in order to increase perfusion pressure in the renal cortex, where most glomeruli are located. Second, AT-II crosses into 2 areas of the brain lacking the blood-brain barrier (the SFO- Subfornical Organ and OVLT – Vascular Organ of the Lamina Terminalis) to trigger the sensation of thirst. Third, AT-II travels to the adrenal cortex, where it stimulates zona glomerulosa cells to release aldosterone.

Aldosterone is the body's main mineralocorticoid. Its site of action is the DCT and collecting ducts of each nephron. As a steroid hormone, aldosterone enters target cells directly, binds to a cytosolic receptor (MR), and translocates into the nucleus, where the hormone-receptor complex alters gene transcription. Aldosterone activates a gene encoding a specialized sodium channel that increases sodium reuptake by 1% or so each time the glomerular filtrate reaches the DCT and collecting ducts. Although a 1% increase in sodium reuptake sounds trivial, bear in mind that each kidney contains 1.5 million nephrons, which collectively filter the equivalent of 800 liters of blood volume each day. This sodium is reabsorbed at the expense of K+ and H+ ions excreted in the urine. This may seem counterproductive until one recognizes that potassium is far more abundant in the environment than sodium. Moreover, multiple metabolic processes yield protons; the challenge is to prevent blood pH from becoming too acidic as opposed to too alkaline.

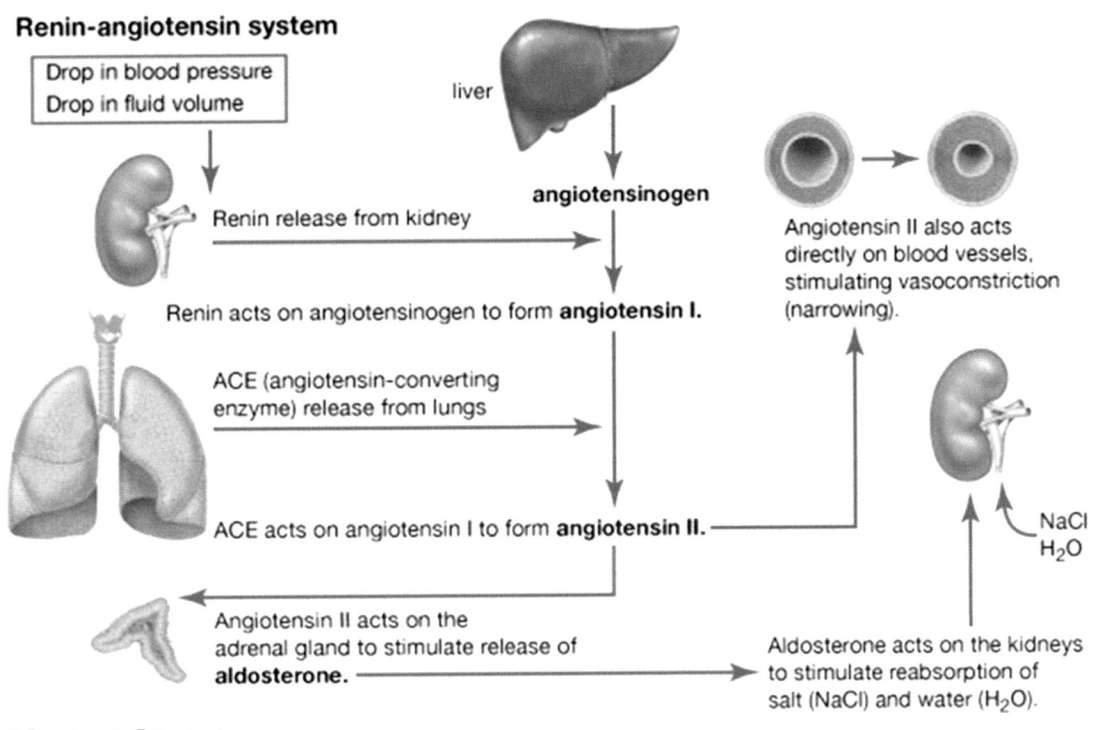

Image courtesy of: https://www.britannica.com/science/renin-angiotensin-system

The diagrams below depict the action of aldosterone at the cellular level.

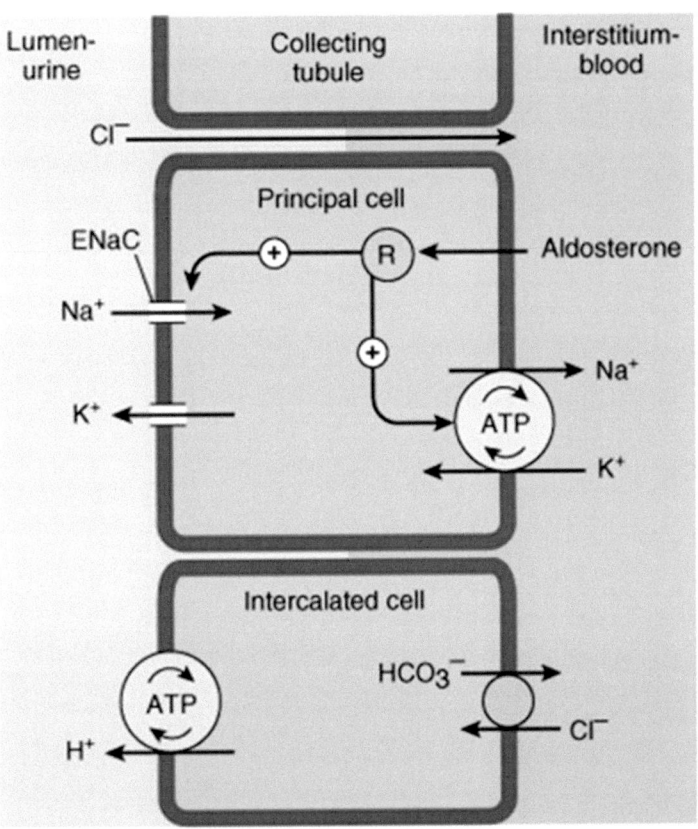

Source: Katzung BG, Masters SB, Trevor AJ: *Basic & Clinical Pharmacology*, 11th Edition: http://www.accessmedicine.com

Copyright © The McGraw-Hill Companies, Inc. All rights reserved.

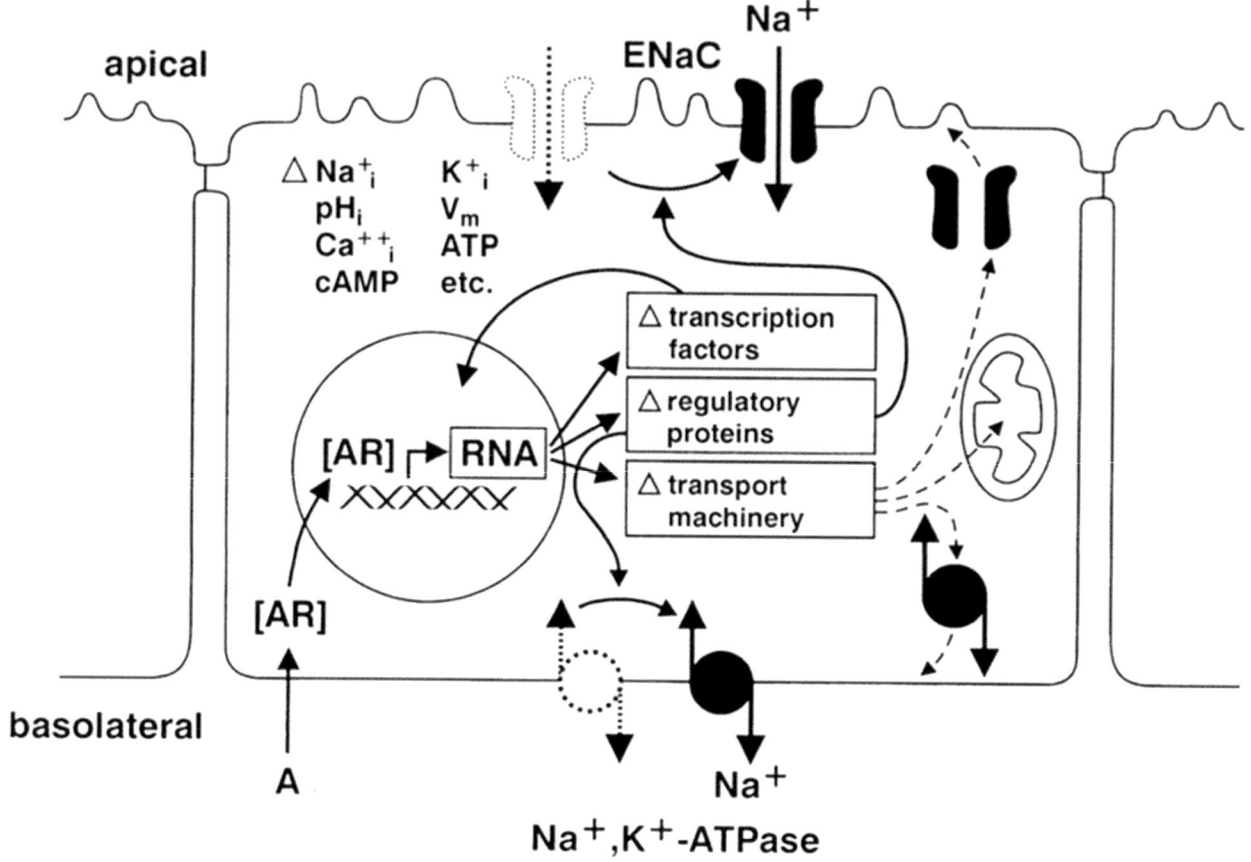

A = Aldosterone AR = Aldosterone/Receptor Complex

Image Courtesy of: http://ajprenal.physiology.org/content/277/3/F319

The kidneys are designed primarily to retain sodium. Passing a sodium load (for example, in a person who drinks seawater) presents the nephrons with a true challenge. The major sodium disposal pathway involves releasing atrial natriuretic peptide (ANP) from the heart's left atrium in response to excessive mechanical stretching (i.e., high blood pressure or volume). ANP binds to guanylate cyclase receptors in the proximal convoluted tubule (the part of the nephron where most sodium reuptake occurs). Guanylate cyclase converts GTP to cyclic GMP; cGMP activates a protein kinase that phosphorylates sodium leak channels, leaving them stuck in the open position for an extended period. In the presence of ANP, less sodium is recovered in the PCT; consequently, more sodium is excreted into the urine.

The diagrams below illustrate the opposing effects of Angiotensin II and ANP.

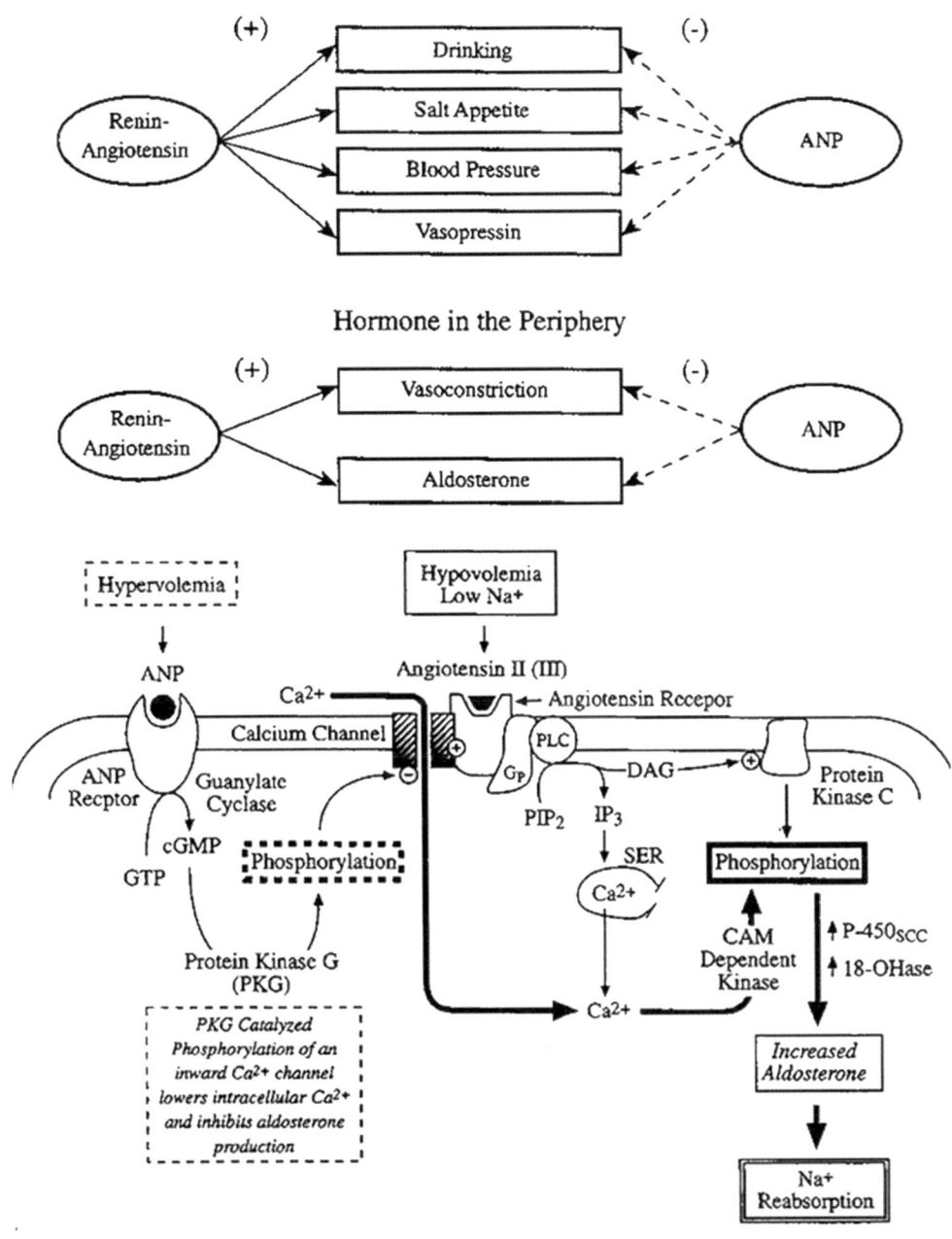

Images Courtesy of:
http://www.nbs.csudh.edu/chemistry/faculty/nsturm/CHE452/22_RenAngioAldoANP18.htm

In spite of the ANP pathway, an excess sodium load leads to water retention due to ADH release from the posterior pituitary (in an attempt to dilute the sodium). The kidneys face the challenge of passing a water load on top of limiting sodium reuptake, which occurs over the span of several days. Ultimately, ingesting saline or seawater requires the kidneys to excrete more water than can be conserved in order to reestablish sodium balance.

Conclusions

Glucose and sodium homeostasis are two vital endocrine and excretory systems activities, respectively. Due to metabolic and environmental factors, the human body has evolved to handle a short-term excess of glucose or sodium far better than an absolute deficiency of either nutrient.

In the case of glucose, the main hormone that lowers blood glucose levels is insulin, the primary signal for anabolism. In contrast, the body has several pathways to raise blood glucose, including glucagon, epinephrine/NE, and stress hormones like cortisol. In the case of sodium, the renin/angiotensin/aldosterone axis conserves sodium at the expense of potassium. Although ANP can increase sodium diuresis, the kidneys must sacrifice water to excrete excess sodium.

Essay 41: The Human Circulatory System

The terms 'circulatory system' and 'cardiovascular system' are often used interchangeably. Although the heart and blood vessels (the cardiovasculature) comprise the core components of the circulatory system, we'll see that several other organs (in particular, the lungs, liver, kidneys, and adrenal glands) play vital roles in the circulatory system's operation.

Most people vaguely know that humans and other mammals possess four-chambered hearts. In physiological terms, it is more useful to think of the heart as a dual pump mechanism. The left side of the heart is the high-pressure side that pumps oxygenated blood throughout the arterial/systemic circulation. In contrast, the right side of the heart is the low-pressure side, whose job is to move deoxygenated venous blood through the pulmonary circulation, where gas exchange occurs.

Before discussing the finer nuances of blood circulation, follow the path of a red blood cell starting at the left side of the heart. Oxygenated blood enters the left atrium (LA) through the relatively short pulmonary veins. From the LA, blood passes through the mitral valve into the left ventricle (LV), the most muscular chamber of the heart. With each contraction, the LV ejects approximately 2/3 of its blood volume into the aorta, the largest artery in the body.

An adult's aorta is roughly the same diameter as a garden hose. Numerous arteries branch off the aorta, starting with the coronary arteries, which provide blood to the heart itself. Next comes the aortic arch, where the brachiocephalic, common carotid and subclavian arteries originate. These arteries provide blood to the head, neck, arms, and much of the spinal column.

Once the aorta descends below the diaphragm into the abdomen, the celiac, superior mesenteric, and inferior mesenteric arteries branch off. These provide arterial blood to the so-called splanchnic circulation, a term encompassing the liver, most of the gastrointestinal tract, and the spleen. Finally, the renal arteries branch above the aortic bifurcation, basically a fork in the road where the aorta gives rise to the arteries of the pelvis and legs.

With a few exceptions, arteries branch into ever smaller vessels called arterioles, which subdivide into capillaries, the smallest of all blood vessels whose walls are a single cell layer thick. This diameter is so small that red blood cells must move through capillaries in a single file line. Tissues throughout the body depend on capillary beds to provide them with oxygen and nutrients and remove carbon dioxide and other waste products.

After blood moves through the capillaries, it begins its return journey to the heart through the venous system as well as other small vessels called lymphatics, which serve as spillways. Small venules merge into veins. Above the diaphragm, venous blood enters the heart's right atrium (RA) through the superior vena cava. Venous blood from structures below the diaphragm flows into the RA through the inferior vena cava. The lymphatic network, present everywhere except the spleen and the brain, transports lymph fluid into the thoracic duct, emptying into the superior vena cava.

From the RA, blood passes through the tricuspid valve into the right ventricle (RV). The RV is far less muscular than the LV. Whereas the LV generates enough force to pump blood several feet, the RV only needs to move blood 6 inches or so through the pulmonary circulation.

The RV pumps blood into the pulmonary artery (PA), the only artery in the body that carries deoxygenated blood. The PA branches numerous times within the lungs until it gives rise to capillary beds surrounding the millions upon millions of air sacs called alveoli. Gas exchange takes place here. After absorbing more oxygen and releasing carbon dioxide, the red blood cells return to the left side of the heart.

Beyond their essential role in gas exchange, the lungs also help regulate blood pressure as part of the renin-angiotensin system, discussed shortly.

Although the heart affects blood pressure by altering its rate and force of contractility, several non-cardiac mechanisms come into play.

First, the liver produces many blood proteins, one of which, albumin, acts as a sponge to minimize fluid leakage out of the blood vessels. The liver also produces angiotensinogen. This protein plays a key role in raising blood pressure. When pressure sensor cells in the kidneys detect low blood flow, the kidneys release an enzyme called renin, which cleaves angiotensinogen into a more active form called angiotensin I. Cells in the kidneys and lungs contain an enzyme called ACE (angiotensin-converting enzyme), which cleaves angiotensin I into its most active form, angiotensin II (AT-II).

AT-II raises blood pressure in three ways: First, it causes arterial smooth muscle to contract, which raises blood pressure in a matter of seconds. Second, AT-II stimulates cells in the adrenal cortex to produce a hormone called aldosterone. Aldosterone is a steroid hormone that increases sodium and water reabsorption in the kidneys. This effectively raises blood pressure by expanding blood volume. Finally, AT-II acts on certain parts of the brain, producing the sensation of thirst. The interplay of the circulatory system with the rest of the body in moving blood and maintaining blood pressure is arguably one of the most complicated yet elegant feats in all of physiology.

Essay 42: Overview of the Human Immune System

It is helpful to divide the immune system into three divisions: non-specific, innate, and adaptive immunity. First, non-specific immune defenses include intact skin, body fluids with an acidic pH (e.g., sweat, gastric juice, and urine), and complement proteins in the bloodstream. Many potentially infectious pathogens are foiled at this level of immunity.

The bulk of this article will address the cellular components of the immune system, which can be subdivided into innate and adaptive immunity.

Most of a person's circulating white blood cells (50-65%) fall into the category of innate immunity, meaning they possess no recognition capacity and respond the exact same way regardless of the nature of the microbial invader. Innate immune cells include granulocytes (neutrophils, eosinophils, basophils, and mast cells) along with monocytes (as well as their tissue counterparts, the macrophages) and NK (natural killer) cells.

Collectively, these cells can be thought of as the shock troops of the immune system. Except for monocytes and NK cells, the other innate immune system cells tend to be short-lived. When they confront foreign cells or antigens, they release a host of chemicals designed to a) combat the invaders directly, b) promote inflammation, and c) blow the proverbial whistle and attract other white blood cells to the site of infection.

These cells are listed in descending order of abundance in the immune system, and their specific functions are described.

1) Neutrophils contain granules with contents like lysozyme and lactoferrin. These proteins attack bacterial cell walls and sequester iron, respectively. Neutrophils also contain enzymes such as myeloperoxidase, which produces hypochlorite ions (a.k.a. bleach), and other enzymes that produce oxygen radicals (nitric oxide, superoxide anions, etc.). These chemicals damage bacteria and surrounding tissue indiscriminately.

Neutrophils respond to cytokines and chemotactic factors such as Interleukin-8, Leukotriene B4, and many other inflammatory mediators.

2) Monocytes/Macrophages are larger than most other white blood cells. They have a distinctive horseshoe-shaped nucleus. These cells are also called phagocytes, meaning they can engulf and destroy bacteria. Macrophages also act as antigen-presenting cells to certain T lymphocytes called T helper cells. This will be discussed in further detail.

3) Natural Killer (NK) cells, formerly called large granular lymphocytes, are believed to be the main cells responsible for immune surveillance. This amounts to patrolling the body, looking for virally infected cells, tumor cells, and pre-cancerous cells.

4) Eosinophils have bi-lobular nuclei and pink staining granules. They produce major basic proteins and other mediators believed to be important in containing parasitic infections, especially worms. Hyperactive eosinophils are believed to contribute to airway damage seen in people with chronic asthma.

5) Basophils and Mast cells figure prominently in allergic reactions, e.g., to pollen and bee stings. These cells contain an abundance of purple staining granules as well as an antibody called IgE on their surface.

When an allergen cross-links two IgE proteins, it triggers a chain of events culminating in the release of histamine, platelet-activating factor (PAF), and other inflammatory molecules from the granules within these cells. Out-of-control histamine release can lead to death from anaphylactic shock.

In contrast to the innate immune system, cells of the adaptive immune system, the lymphocytes, are capable of specific antigen recognition and possess the ability to respond more quickly and efficiently if they encounter the same antigen again. The adaptive immune system is traditionally divided into two arms: humoral immunity, dominated by B lymphocytes, and cell-mediated immunity, dominated by T lymphocytes.

Unlike neutrophils or monocytes, a microscope cannot distinguish lymphocytes as B cells or T cells. Instead, a flow cytometer is required to sort lymphocytes based on differences in their surface proteins.

B lymphocytes originate and mature in the bone marrow. These cells produce immunoglobulins, better known as antibodies.

Antibody proteins may be soluble, like gamma globulin or antibodies in breast milk, or be surface-bound, like IgE. B cells can generate hundreds of thousands of different antibodies. The basis of this tremendous diversity, a mechanism called gene rearrangement, was not discovered until 1976.

In the aftermath of an infection, a subset of B cells that produced the most effective antibodies persist in the spleen and lymph nodes as memory B cells.

T cells originate in the bone marrow and then migrate to the thymus gland, where they undergo maturation and selection. Rather than secrete antibodies, T cells produce a diverse array of surface proteins called T cell receptors (TCR). As with B cells, the mechanism underlying TCR diversity involves gene rearrangement.

Several categories of T cells exist, including T helper cells (CD4 positive T cells), cytotoxic T cells (CD8 positive T cells), and suppressor T cells.

T helper cells process reports from antigen-presenting cells, which include macrophages, B cells, and follicular dendritic cells. TH1 cells interact with macrophages, whereas TH2 cells mainly interact with B cells. In general, T helper cells communicate with antigen-presenting cells, cytotoxic T cells, and each other by means of cytokines, esp. Interleukin-2.

Cytotoxic T cells (CTL) specialize in destroying virally infected cells by direct cell-to-cell killing. The mechanism of cell killing remains obscure but probably includes producing oxygen radicals, releasing inflammatory mediators, complement fixation, and induction of apoptosis (programmed cell death).

Suppressor T cells are believed to contain inflammation at the site of infection and shut down the immune response once the invaders have been repelled. Several cytokines appear to be involved in turning off inflammation, esp. Interleukin-10 and TGFβ.

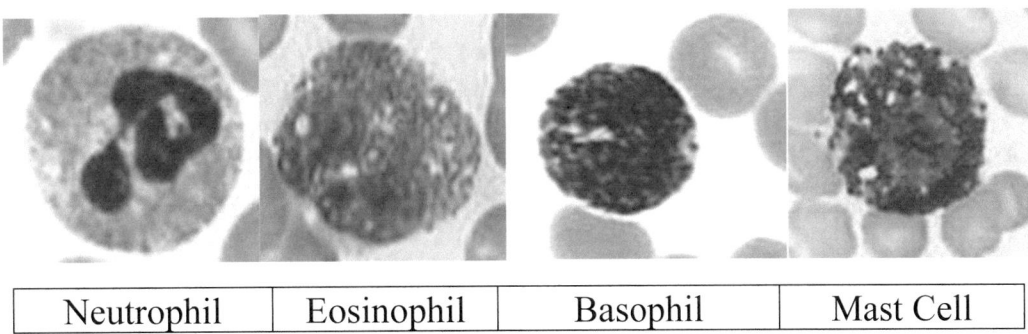

| Neutrophil | Eosinophil | Basophil | Mast Cell |

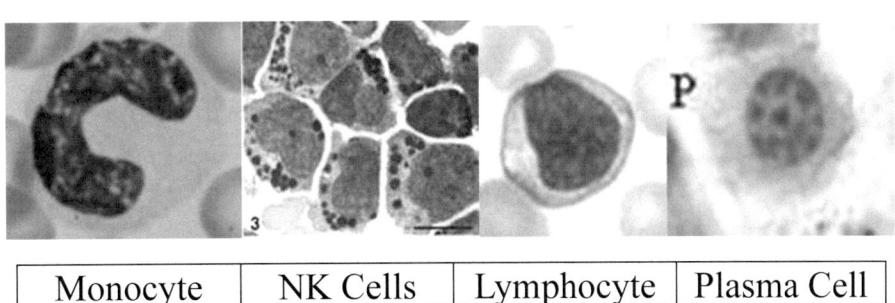

| Monocyte | NK Cells | Lymphocyte | Plasma Cell |

Essay 43: How do human bones age?

As with other organ systems, the human skeleton ages in a predictable way. Over time, the cortical layer of bone (most prominent in the long bones of the appendages) turns over more slowly. This is probably due to a decline in the number and activity of the cells responsible for remodeling the bone cortex. Consequently, bone fractures take longer to heal in elderly people.

The cells responsible for bone remodeling, known as osteoblasts and osteoclasts, play a central role in the disease osteoporosis, which may affect over 40 million people in the U.S. alone. Although this disease is not completely understood, it seems to result from an imbalance in the rate of calcium deposition by osteoblasts compared to bone resorption by osteoclasts. Some factors in this imbalance may be a drop in estrogen production in menopausal women as well as decreased production of vitamin D in both sexes with advancing age. Low vitamin D levels translate into less dietary calcium absorbed from the gut, which means less calcium is available for storage in the bones. It is important to remember that osteoporosis occurs only in living bone tissue; dead bones cannot become osteoporotic.

Most cases of osteoporosis occur in middle-aged to elderly women. Postmenopausal Caucasian and Asian women seem to be the most at risk of all female populations. Women who are of normal weight or obese are somewhat less likely to develop osteoporosis, possibly because of higher levels of estrogen stored in adipose tissue.

Men occasionally develop osteoporosis; however, several factors work in a man's favor to decrease the risk. First, men tend to achieve a higher peak bone density than women. Second, most women experience one or more pregnancies during their lives, each time depleting calcium reserves in their bones. Finally, men produce testosterone throughout their lives, and this hormone (along with estrogen produced as a metabolite) appears to be somewhat protective against osteoporosis.

Most cases of osteoporosis are primary, meaning no underlying cause precipitated the disease. Most secondary cases of osteoporosis result from endocrine disorders, especially hyperparathyroidism, a disease state marked by the excessive release of PTH (parathormone) from the parathyroid glands. Other cases are caused by chronically elevated levels of the hormone cortisol.

Hypercortisolemia may be due to adrenal or pituitary tumors (Cushing's Syndrome) but more often results from long-term treatment with high doses of exogenous corticosteroids (for conditions like asthma or lupus). Yet another uncommon cause of secondary osteoporosis is untreated hyperthyroidism.

In addition to changes in the bone cortex, the bone marrow also changes with age. By adulthood, much of the red marrow inside of one's long bones is replaced by fatty white marrow as blood cell production plateaus. By old age, this transformation is complete. Because some level of blood cell production is necessary to sustain life, this process continues in the marrow of the axial skeleton, i.e., the sternum, ribs, vertebrae, and pelvic bones.

First line treatment for osteoporosis consists of oral bisphosphonates, especially Alendronate (trade name Fosamax). These drugs are thought to work by stabilizing calcium phosphate salts in the bone cortex, making them more resistant to alkaline phosphatase and other osteolytic activity by osteoclasts. Patients who do not tolerate oral medications (due to severe acid reflux) can try an infusion of Zoledronic Acid (Zometa) at 90-day intervals.

Beyond the bisphosphonates, resistant osteoporosis sometimes responds to the parathormone (PTH) analog teriparatide, better known by its trade name Forteo. This drug mimics PTH but with minimal osteoclast activation. Forteo dramatically reduces PTH release from the parathyroid glands. The downside is that Forteo must be injected daily and can be used for a maximum of 2 years.

Essay 44: Mitochondria

Mitochondria are organelles found exclusively in eukaryotic cells, meaning protozoa, fungi, plants, and animals. The term 'mitochondrion' is derived from a Greek word meaning thread. This accurately describes their appearance in the light microscope as barely visible thread-like structures. Following the invention of the electron microscope, scientists learned that mitochondria have a complex structure that allows them to harness metabolic energy in a useful form.

As an aside, prokaryotic cells (eubacteria, blue-green algae, and archebacteria) lack mitochondria but maintain a strong evolutionary kinship with them. Biologists believe that free-living, aerobic prokaryotes gave rise to mitochondria over 1 billion years ago when they were captured by the ancestors of today's eukaryotic cells.

Thus, mitochondria (as well as chloroplasts in plant cells) can be thought of as obligate endosymbionts, as opposed to organelles whose synthesis is directed entirely by the nucleus. Unlike ribosomes, ER, Golgi, or lysosomes, mitochondria contain their own DNA and ribosomes and reproduce in a manner similar to bacteria. Over time, however, many of the genes encoding mitochondrial proteins have been "captured" and integrated into the cell's nuclear DNA, reducing mitochondria to a semiautonomous status.

The powerhouse of the cell

Far and away, mitochondria's primary role is fuel molecule metabolism. Because the entire structure of the mitochondria is optimized for the production of a proton gradient, a more accurate description of these organelles would be "the cell's batteries." Specifically, mitochondria are the site of aerobic respiration, where sugars and fatty acids are broken down in the presence of oxygen to produce energy in the form of ATP. Carbon dioxide is formed as a byproduct. As we shall see, far more ATP is generated by the mitochondria than by the cytosolic enzymes responsible for anaerobic glycolysis.

Glycolysis, or anaerobic respiration, is the first phase of sugar breakdown at the cellular level. Simple sugars such as glucose, fructose, and galactose are converted to a three-carbon compound called pyruvate. Because this process occurs in the cytosol and does not require oxygen, it is termed anaerobic. The net energy yield from anaerobic respiration is two molecules of ATP.

In the absence of oxygen, pyruvate is converted to lactic acid, which is a metabolic dead end. Although the liver can convert lactic acid to pyruvate and pyruvate back into glucose, this requires the hydrolysis of more ATP.

Unicellular organisms like yeast and most bacteria can survive by anaerobic respiration alone, but human cells cannot. In the presence of oxygen, pyruvate enters the mitochondria), where it is broken down by an enzyme complex called PDH (pyruvate dehydrogenase).

PDH catalyzes an irreversible reaction in which pyruvate is converted to carbon dioxide and a two-carbon acetate bound to a carrier molecule called coenzyme A, or acetyl CoA for short. Fatty acids are broken down into acetyl CoA by a process called beta-oxidation, which also requires oxygen.

The Krebs Cycle and Electron Transport Chain

Regardless of its fuel molecule of origin, acetyl CoA enters the mitochondrial matrix, where the enzymes of the Krebs cycle harvest its energy to generate the electron carriers NADH and FADH2. These high-energy electron carriers, both derived from B vitamins, release electrons, which are then shuttled through a series of iron-containing proteins called cytochromes found along the inner mitochondrial membrane.

The flow of electrons is absolutely dependent on the presence of oxygen, which serves as the final electron acceptor. Without oxygen, this electron flow, the Krebs cycle, and acetyl CoA production all grind to a halt.

The cytochromes and their associated cofactors are collectively known as the electron transport chain, or ETC for short. The ETC couples the flow of high-energy electrons to transport protons into the space between the inner and outer mitochondrial membranes. The resulting proton gradient drives the synthesis of ATP as protons flow back into the mitochondrial matrix through a protein complex called ATP synthetase. Thirty-six molecules of ATP are produced for every molecule of glucose metabolized aerobically. This represents an 18-fold increase in energy yield compared to the ATP produced by anaerobic respiration.

In a sense, the mitochondria act as microscopic proton batteries charged by the breakdown of fuel molecules into carbon dioxide and water. This process can also be viewed as a form of enzymatic combustion, in which carbon-rich fuel molecules are burned in the presence of oxygen to yield useful chemical energy in the form of ATP.

Steroid hormone synthesis

Mitochondria are also responsible for the initial step of steroid hormone production. Some contain an enzyme called CYP450 SCC (side chain cleavage), which converts cholesterol to pregnenolone. This specialized function is restricted to mitochondria in the adrenal cortex, testes, and ovaries. Subsequent steroid hormone synthesis steps are catalyzed by enzymes in the smooth endoplasmic reticulum (sER).

Heat production

Finally, mitochondria in specialized adipose tissue called brown fat generate heat by short-circuiting the electron transport chain. As in other cells, these mitochondria burn fatty acids aerobically. The major difference is these mitochondria contain a specialized protein called thermogenin, which allows protons to flow from the intermembrane space back into the mitochondrial matrix without having to pass through the ATP synthetase complex. As such, these mitochondria produce heat at the expense of ATP production. However, this process is not wasteful; the heat generated by brown fat ensures that arterial blood remains liquid even at low temperatures. By extension, lipolysis of brown fat helps maintain core body temperature.

Essay 45: Neurotransmitters

A neurotransmitter (NT) is a chemical messenger released from a neuron's synaptic terminal following an action potential. Until the middle of the 20th century, some scientists thought the human nervous system required as few as two neurotransmitters to function - one substance to excite neurons and a second to inhibit them. This model of neuronal activity, analogous to a radio or TV set with a single On/Off button, proved to be extremely oversimplified. Today, scientists classify over 100 different molecules as neurotransmitters. At first glance, the sheer number of NTs seems daunting; however, the vast majority of them fall into three broad categories: excitatory NTs, inhibitory NTs, and modulatory NTs.

Excitatory Neurotransmitters

In the central nervous system (CNS), the amino acids glutamate and aspartate serve as the major excitatory NTs. These NTs work by opening ligand-gated ion channels on their target neuron, allowing extracellular sodium (as well as calcium) to flow into the cell. Most neurons have a resting membrane potential of approximately -70 mV (millivolts). When sufficient ion channels open, the cumulative influx of these cations depolarizes the cell membrane above the threshold voltage (around +30 mV), and the neuron fires an action potential. Glutamate is especially prominent in the cerebral cortex and hippocampus, where many synapses are excitatory.

At the neuromuscular junction, motor neurons release acetylcholine, which triggers skeletal muscle contraction. As it turns out, acetylcholine is one of the most versatile NTs, with distinct functions in the heart, CNS, and parasympathetic nervous system. Depending upon the type of membrane receptor it binds to, acetylcholine can stimulate gut motility, constrict the pupils, decrease heart rate, as well as play a role in learning and memory at the CNS level.

Inhibitory Neurotransmitters

In the CNS, the amino acids GABA (gamma amino butyric acid) and glycine are the main inhibitory NTs. These NTs open anion channels, allowing chloride ions to enter the neuron. The anion influx hyperpolarizes the cell's membrane potential to -90 mV. This temporarily prevents the neuron from firing action potentials; over time, the cell uses various ion pumps to reset the membrane potential to its resting state.

In cardiac muscle, acetylcholine acts as an inhibitory neurotransmitter. It slows heart rate by causing a G protein-gated potassium channel to open, leading to an efflux of potassium ions from cardiac conduction cells. The outflow of potassium ions results in membrane hyperpolarization, effectively slowing heart rate.

Modulatory Neurotransmitters

The largest group of NTs is the so-called modulatory transmitters or neuromodulators. Their function is to enhance or blunt the effects of other NTs. Returning to the analogy of a TV set, neuromodulators correspond to the contrast and volume knobs, which allow you to adjust the appearance of images and intensity of sounds rather than simply turn the television on or off.

Neuromodulators include amino acid derivatives such as dopamine, norepinephrine, serotonin, melatonin, and histamine. These monoamines play diverse roles in sleep, arousal, attention, and mood. Acetylcholine and GABA act as neuromodulators at some CNS synapses, including the basal ganglia and hippocampus, which are involved in speech and memory, respectively.

Next, an assortment of peptides modulates neuronal activity. Examples of these peptides include enkephalins, endorphins, and dynorphins, all of which decrease the sensation of pain; vasopressin, which enhances thirst; ghrelin and orexin, which stimulate appetite; and Substance P, which seems to enhance pain signals in the periphery and perhaps in the spinal cord as well.

Last but not least, a host of substances not traditionally considered to be neurotransmitters seem to play a role in neuronal signaling. These range from nucleotides like ATP and the purine molecule adenosine to membrane lipids such as arachidonic acid to soluble gases like nitric oxide.

Essay 46: Pituitary Gland Function

The pituitary gland is a pea-sized region of specialized endocrine cells and neurons located behind the optic chiasm and enclosed in a bony structure called the sella turcica, or Turkish saddle. Although the pituitary is often called the "master gland" of the endocrine system, that label is more appropriate for an adjacent area of the brain known as the hypothalamus. As 19th-century scientists deciphered the anatomy and physiology of the central nervous system and endocrine glands, the role of the pituitary gradually became clear.

Today, biologists separate pituitary function into the anterior pituitary (adenohypophysis) and posterior pituitary (neurohypophysis). Each will be considered in turn.

The anterior pituitary is connected to the hypothalamus by a specialized network of blood vessels known as the hypophyseal portal system. Hypothalamic neurons synthesize proteins called release factors, which signal various populations of pituitary cells to release their hormones directly into the systemic circulation. The major cell types in the anterior pituitary include somatotrophs, lactotrophs, thyrotrophs, gonadotrophs, and corticotrophs.

Somatotrophs produce hGH or human growth hormone. Growth hormone acts in the liver to stimulate the production of IGF-1 (Insulin-Like Growth Factor-1). IGF-1, in turn, promotes bone and muscle growth prior to puberty. Growth hormone production declines after puberty; its importance in adult physiology remains a matter of debate.

Lactotrophs make a protein hormone called prolactin (Prl). Prolactin stimulates breast development and milk production in females starting in the last trimester of pregnancy. The importance of prolactin in male physiology remains unclear. Males do not lactate under normal circumstances. One exception is a patient with a pituitary tumor called a prolactinoma, whose symptoms include abnormal milk production, also called galactorrhea.

Thyrotrophs produce thyroid-stimulating hormone (TSH), which stimulates follicular cells in the thyroid gland to synthesize thyroid hormones. These hormones called T3 and T4, require iodine atoms to be functional.

Humans who live in mountainous regions far from the ocean are most at risk of iodine deficiency. The hallmark of an iodine-deficient diet is goiter, which results from an enlarged, swollen thyroid gland.

As a general rule, hypothyroid states (like iodine deficiency) are marked by high TSH levels, whereas hyperthyroid states (e.g., Grave's disease) are marked by low TSH levels. These phenomena are direct results of feedback inhibition, or loss thereof, by thyroid hormones in the pituitary gland. Excessive levels of T3 and T4 shut off TSH production. Conversely, a lack of thyroid hormone production leads to enhanced TSH release by the pituitary. Similar trends occur with imbalances of other endocrine hormones, most notably cortisol, estrogen, and testosterone.

Gonadotrophs make the hormones FSH (follicle-stimulating hormone) and LH (luteinizing hormone). In females, an FSH spike occurs at the beginning of the menstrual cycle. FSH triggers the development of egg follicles and stimulates estrogen production by the ovaries. An LH spike triggers ovulation midway through the menstrual cycle. In males, FSH stimulates sperm production and maturation in the seminiferous tubules of the testes. LH acts in a complementary manner by stimulating testosterone production in specialized testicular cells called Leydig cells.

Corticotrophs produce POMC (pro-opiomelanocortin), a precursor hormone that is cleaved to yield β endorphin, ACTH (adrenocorticotropic hormone), MSH (melanocyte-stimulating hormone), and some minor peptides. ACTH triggers the production of the steroid hormone cortisol in the adrenal cortex. Endorphins blunt the transmission of pain signals in the spinal cord. MSH promotes the synthesis of the pigment melanin in the skin, hair, and iris of the eye.

Posterior pituitary

The posterior lobe of the pituitary gland does not make its own hormones. Instead, axons from two groups of hypothalamic neurons - the supraoptic nucleus (SON) and paraventricular nucleus (PVN) – terminate in the posterior pituitary.

These specialized neurons produce the hormones ADH (antidiuretic hormone), also known as vasopressin and oxytocin. When a person becomes dehydrated, osmoreceptors in the brain trigger ADH release into the systemic circulation. ADH travels to the kidneys, promoting water reuptake in the epithelial cells lining the collecting ducts.

ADH's exact mechanism of action remained obscure until 1990 when Peter Agre discovered a class of protein channels, now called aquaporins, which selectively allow water molecules to cross the cell membrane. ADH activates a G-protein coupled receptor on these epithelial cells, triggering an influx of calcium ions, the activation of Protein Kinase C, and the translocation of aquaporins to the cell surface, leading to enhanced water reuptake.

Oxytocin has two main functions. First, it triggers uterine contractions when a pregnant woman goes into labor. Second, it promotes the movement of milk from the breast ducts to the nipple by stimulating the contraction of myoepithelial cells lining the ducts. In lactating women, this is sometimes called the milk-let-down reflex. The function of oxytocin in male physiology remains uncertain.

Essay 47: Platelets

The formed elements of human blood include red blood cells, white blood cells, and platelets. The first two cell types arise individually from myeloid and lymphoid stem cells in the bone marrow. In contrast, the source of platelets is a distinct precursor cell called the megakaryocyte, which literally means giant bone cell. Megakaryocytes are among the largest cells in the human body. Unlike other cell types, megakaryocytes contain a polyploid nucleus, meaning their DNA has been replicated more than once; in this case, ten or more rounds of DNA replication occur during megakaryocyte development.

During its lifetime, each megakaryocyte gives rise to millions of platelets cleaved from the parent cell as cytosolic fragments lacking DNA or organelles. Red blood cells also lack a nucleus at maturity but live approximately 120 days. Platelets last 7 to 10 days on average. As with red blood cells, the liver and spleen remove senescent or non-functional platelets from the circulation.

The major function of platelets is to help form blood clots, especially in the arterial blood vessels. When the wall of a blood vessel is damaged or endothelial cells are sheared off the basement membrane, the collagen-rich extracellular matrix (ECM) is exposed. Circulating platelets bind to an ECM protein called Von Willebrand Factor via a surface receptor called GPIb. Platelets form a logjam to plug small tears and holes directly.

When the damage is more extensive, a series of enzymes known as the coagulation cascade is activated. This enzymatic cascade culminates in the formation of insoluble fibrin strands that trap platelets and other blood cells inside a web-like mesh, ultimately forming a blood clot or thrombus.

Platelets are also recruited to areas of inflammation by a variety of soluble factors. These include PAF (platelet-activating factor) released by basophils; TXA2 (Thromboxane A2), an arachidonic acid derivative released from a variety of inflammatory cells; serotonin, released from platelets themselves; and adenosine, which also triggers platelet aggregation.

Activated platelets become sticky by expressing higher levels of adhesion molecules on their surfaces. In addition to GPIb, which promotes adhesion to blood vessel walls, platelets also express the surface glycoprotein GPIIb/IIIa, which allows them to bind to each other more readily.

Blood clot formation is necessary for survival but can be life-threatening when it happens inappropriately. Anticoagulants like Coumadin (warfarin) target the clotting factors produced in the liver. In contrast, antiplatelet drugs like aspirin, dipyridamole, and clopidogrel (Plavix) work by blocking platelet aggregation. Newer antiplatelet drugs include monoclonal antibodies and other antagonists against the GPIIb/IIIa receptor. These agents, including Abciximab (ReoPro) and Eptifibatide (Integrilin), are mainly used in the hospital setting.

Essay 48: How are Proteins Converted to Amino Acids? It's all about Proteases

In humans and most other animals, dietary proteins cannot be absorbed in an intact form. They must be broken down into amino acids by enzymes secreted in pancreatic juice as well as those within the enterocytes lining the small intestine. The enzymes that accomplish this task are collectively referred to as peptidases or proteases. This article will focus on the enzymatic breakdown of proteins instead of proteolysis performed by inorganic chemical methods.

Before discussing proteases themselves, it is important to understand a bit about protein architecture as well as the chemical structure of the peptide bond. Biochemists describe proteins in terms of a four-tiered structural hierarchy. A protein's primary structure consists of its amino acid sequence from the initial N-terminus to the C-terminus and any disulfide bonds between cysteine residues. Although a protein's amino acid sequence is the fundamental determinant of its three-dimensional structure, scientists are often unable to predict a protein's 3D conformation on the basis of primary structure alone.

The secondary protein structure consists of defined, localized shapes: alpha helices, beta-pleated sheets, beta strands, and hairpin turns. In some proteins, these elements form super-secondary structures called motifs, each with distinct functions. For example, the EF-hand motif of calmodulin binds calcium ions, whereas the helix-loop-helix motif in many transcription factors recognizes DNA. Motifs are themselves organized into higher-order structures called domains, which may contain the catalytically active site of an enzyme or the binding region of a globular protein.

Tertiary structure refers to the overall three-dimensional shape of a protein. It is often difficult to draw a sharp line between domains and tertiary structure, as some proteins consist of little more than a single domain. In other proteins, amino acids from distant parts of the primary sequence are folded together to form the active site. At any rate, this is the highest level of architecture in monomeric (single subunit) proteins. In contrast, proteins consisting of multiple subunits, such as hemoglobin, are said to exhibit a quaternary structure.

In spite of the almost endless variety of proteins in nature, their common denominator is a backbone formed by peptide bonds. Organic chemists describe the peptide bond as an amide, which is a planar unit consisting of a carbonyl group (C=O) whose carbon atom is bonded to the nitrogen atom of an NH group (also known as a secondary amine). Under acidic or alkaline conditions, this amide can be split by hydrolysis (which literally means destruction by water). Hydrolysis of a peptide bond yields a carboxyl group (COO-) at one end and an NH2 group at the other end. Successive hydrolysis reactions can reduce even the largest proteins to individual amino acids.

Biochemists have identified four major families of proteases, all capable of hydrolyzing peptide bonds: serine proteases, zinc proteases, thiol proteases, and acid proteases.

Serine proteases

As the name suggests, these enzymes contain a serine residue at their active sites, along with histidine and aspartate, forming a so-called catalytic triad. The most well-studied members of this family include the digestive enzymes trypsin, chymotrypsin, and elastase, as well as the clotting factors responsible for blood coagulation. Serine proteases are selective in which peptide bonds they cleave.

Chymotrypsin, for instance, hydrolyzes the C-terminal ends of the amino acids phenylalanine, tyrosine, tryptophan, and methionine. Digestive enzymes and clotting proteins must be stored as inactive precursors called zymogens, or else they would destroy the intestinal lining or cause blood to clot inappropriately.

Zinc proteases

These proteases contain a zinc ion as a cofactor in their active sites. As with serine proteases, they tend to hydrolyze amino acids with aromatic or bulky side chains. Important zinc proteases in humans include carboxypeptidase A, collagenase, and matrix metalloproteases.

Thiol proteases

Thiol proteases contain a cysteine residue in their active sites. One example is papain, an enzyme found in papaya juice.

Acid proteases

Acid proteases require a low pH to become activated. In humans, they include the digestive enzyme pepsin, secreted by gastric chief cells, as well as several lysosomal enzymes.

Essay 49: How Do Steroids Work in the Body?

Before elaborating upon this question, it is important to note that the term 'steroids' is not synonymous with the endogenous steroid hormones produced in the adrenal glands and gonads. Endogenous steroids are produced in smaller quantities than their synthetic counterparts. Their production is also self-limited due to their negative feedback effects on the pituitary gland and hypothalamus. Nevertheless, both kinds of steroids exhibit the same mechanisms of action, which will be this article's subject.

General Mechanism of Action

Both endogenous and synthetic steroids are cholesterol derivatives; as hydrophobic lipid molecules, they can cross cell membranes without having to bind to a specific surface receptor. Once inside the cell, steroids bind to their respective protein receptors in the cytoplasm or cell nucleus. When the steroid-receptor complex enters the nucleus, it binds to specific DNA sequences, altering gene expression.

The steroid-receptor complex can turn several genes on or off over the course of minutes, hours, or days. Eventually, the steroid molecules diffuse out of the cell and back into the bloodstream, where they are metabolized in the liver and excreted in the urine. Since this step occurs relatively slowly (especially for synthetic steroids), steroid metabolites can be detected in a urine sample weeks or months after the steroid was administered.

Steroids can be divided into three groups: glucocorticoids, mineralocorticoids, and sex hormones (androgens, estrogens, and progesterone). Each category is discussed below.

Glucocorticoids

This class of steroid includes hydrocortisone, dexamethasone, prednisone, and many others. They all mimic the actions of cortisol, a major stress hormone, in terms of its anti-inflammatory effects. Glucocorticoids are used to treat a variety of conditions ranging from skin rashes and nasal allergies to severe diseases like asthma, lupus, and rheumatoid arthritis.

When used for 14 days or longer, a high-dose steroid cannot be discontinued abruptly but instead must be tapered over several weeks. The reason is that chronically elevated levels of glucocorticoids in the bloodstream suppress cortisol output from the adrenal glands.

Abruptly stopping a glucocorticoid does not give the pituitary time to reactivate cortisol production in the adrenal cortex. The result is a medical emergency known as the Addisonian crisis.

When used chronically, systemic glucocorticoids produce many adverse side effects, including weight gain, diabetes, high blood pressure, hair growth, osteoporosis, stretch marks, easy bruising, and vulnerability to infections. This constellation of signs and symptoms is known in the medical literature as Cushing's syndrome.

Mineralocorticoids

This type of steroid mimics the hormone aldosterone, which increases sodium retention in the kidneys and helps maintain normal blood pressure. The main synthetic mineralocorticoid used medically is called Fludrocortisone (Florinef).

This drug is given to patients diagnosed with adrenal failure, e.g., Addison's disease, or other endocrine disorders such as congenital adrenal hyperplasia. Fludrocortisone may be used chronically to replace aldosterone without producing adverse effects.

Androgens, Estrogens, and Progesterone

Natural androgens include testosterone and dihydrotestosterone. These hormones are largely responsible for secondary male sexual characteristics such as facial hair and a deep voice. High doses of androgens, better known as anabolic steroids, increase muscle mass over the course of several months. This phenomenon accounts for the rampant use of steroids among professional athletes, especially weightlifters, American football players, and pro wrestlers.

Unfortunately, chronic use of anabolic steroids results in damage to the liver, heart, and possibly the brain. Another adverse effect is testicular atrophy (and sometimes male infertility) due to the testes' suppression of endogenous testosterone production.

In the past, testosterone was used to treat refractory anemia, especially in men. Today, it has been largely replaced by erythropoietin and other stimulators of red blood cell production.

Estrogen plays a key role in the menstrual cycle and helps maintain secondary female sexual characteristics. Various estrogens are often used in hormone replacement therapy to relieve post-menopausal symptoms such as hot flashes. In the past, unopposed estrogens were used in oral contraceptive pills (OCP); however, it was later discovered that combining estrogen with progesterone dramatically reduced the risk of breast and uterine cancer.

Progesterone, produced by the corpus luteum and placenta, is the main hormone involved in maintaining a woman's uterine lining during pregnancy. Medically, progesterone is most often used in combination with estrogens in OCPs and hormone replacement therapy. If a pregnant woman must undergo surgery, a progesterone injection is given postoperatively to minimize the chances of preterm labor or miscarriage.

Essay 50: Temperature Control Mechanisms of the Human Body

The temperature of the human body is regulated mainly by the hypothalamus. This small brain region serves as the interface between the nervous and endocrine systems. One of its many tasks is to maintain body temperature within a relatively narrow range - between 95 and 100 degrees Fahrenheit. Below 95o F, a person rapidly develops hypothermia; above 100o F, a person becomes febrile and lethargic. Fevers of 105o F or higher can lead to brain damage or even death if not brought under control. Most of the time, however, body temperature is maintained within this margin of safety.

Since the causes of hypothermia are usually obvious, for example, prolonged exposure to freezing temperatures, this article will focus on fluctuations of body temperature within the normal range as well as the main causes of fever.

Temperature as a circadian rhythm - Body temperature rises and falls over a period of about 24 hours. Several metabolic parameters, such as blood pressure, heart rate, and cortisol levels, exhibit similar circadian cycles. Body temperature is generally lowest in the morning immediately after awakening, probably because metabolism is slower during sleep than at any other time of the day. Body temperature rises throughout the day by one to two degrees Fahrenheit.

Physical exertion, such as intense workouts, may raise body temperature to 1000 F but seldom higher. Strenuous physical activity at room temperature rarely results in fever because the hypothalamus maintains an established set point for body temperature, much like a thermostat. When the hypothalamic neurons and other thermoreceptors detect a rise in core body temperature, compensatory mechanisms, most notably sweating, dissipate the excess heat. Sweat glands are innervated by the sympathetic nervous system, the branch of the autonomic nervous system designed to execute the fight or flight response. Sweating occurs almost exclusively as an involuntary reflex; it is common knowledge that strong emotions like arousal, fear, and anger can also trigger sweating.

Fever - Although people have associated fever with illness since time immemorial, the mechanisms underlying the febrile response remain incompletely understood.

In many cases of infection, antigens from the infectious agents and the immune system's inflammatory response combine to produce fever. Interferons and cytokines like Interleukin-1 can induce fever and pyrogenic bacterial toxins. Viral infections like influenza or chicken pox can also cause fever; on the other hand, self-limited viruses like the adenovirus strains responsible for the common cold seldom cause fever.

In addition to infectious agents, fever has many other etiologies. The important ones include drug reactions, autoimmune diseases (systemic lupus or rheumatoid arthritis), and the most dreaded diagnosis of all, cancer, especially leukemia and lymphoma. The common thread in all of these disease states seems to be out-of-control immunological activity. In contrast, immune-compromised patients, especially those with AIDS or advanced-stage cancer, may remain afebrile even during an overwhelming infection, i.e., septic shock.

Drug reactions deserve special mention. Many adverse drug reactions occur idiosyncratically, meaning sporadic and unpredictable. This is often the case with a febrile response to antibiotics as well as certain other medications. Anesthetic agents like halothane can trigger a condition called malignant hyperthermia in susceptible patients.

Antipsychotic drugs with similar chemical structures, such as haloperidol, can induce a closely related condition called neuroleptic malignant syndrome (NMS), marked by tremors, muscle rigidity, and high fever (sometimes called hyperpyrexia).

The underlying cause of malignant hyperthermia (and possibly NMS) seems to be an abnormal genetic variant of a calcium channel in skeletal muscle called the dihydropyridine receptor. For unclear reasons, halogenated anesthetic and neuroleptic agents cause excessive calcium influx through this channel, resulting in prolonged muscle contractions and subsequent fever. The treatment of choice for NMS is a muscle relaxant such as dantrolene.

Essay 51: Vitamin K and the Coagulation Cascade

The human blood clotting cascade consists of a tightly regulated network of enzymes designed to contain damage to the blood vessels by means of coagulation, thus restoring vascular integrity. The clotting cascade swings into action when a blood vessel suffers penetrating trauma, blunt force trauma, or an internal rupture. The various proteins activate one another, culminating in a clot composed of a platelet plug and cross-linked strands of fibrin. The clot, or thrombus, acts as damage control, stopping further loss of blood through the wound.

Physiologists divide the clotting cascade into three branches: the extrinsic pathway, consisting of Factors III and VII; the intrinsic pathway, consisting of Factors XII, XI, IX, and VIII; and the final common pathway, consisting of Factors X, V, prothrombin (II), fibrinogen (I) and factor XIII. Because the clotting factors were numbered in the order of their discovery, they do not follow a sequential order. (Factor IV was later discovered to be calcium, and Factor VI was omitted as a redundancy after two labs discovered the same enzyme independently).

Far and away, the most common disease involving a protein in the clotting cascade is Factor VIII deficiency, better known as Hemophilia A. An estimated 30,000 males in the U.S. are believed to have Factor VIII deficiency. Because the gene responsible for this disease is located on the X chromosome, the pattern of heredity most often observed is a female carrier giving birth to a son with hemophilia. (Males inherit a Y chromosome from their father and a single X chromosome from their mother).

Females rarely develop hemophilia. For this to occur, the girl's father would have to have hemophilia himself, and her mother would have to carry the defective gene. Hemophilia A is treated with injections of recombinant Factor VIII. (The term recombinant denotes any protein produced by genetically engineered bacteria.) In emergencies, fresh frozen plasma or cryoprecipitate can be used as a substitute for Factor VIII.

Factor IX deficiency, also known as Hemophilia B, is an X-linked disorder. It occurs about 10% as often as Hemophilia A, and affects an estimated 3,300 males in the U.S. Most cases of Factor IX deficiency come to medical attention when a patient with hemophilia fails to respond to injections of Factor VIII. Hemophilia B is treated with injections of recombinant Factor IX.

Some cases of Factor XI and XII deficiency have been reported; however, these disorders are quite rare and tend to be milder than Hemophilia A or B.

Another clotting disorder is Von Willebrand disease. This disorder is caused by mutations or deletions of the gene encoding a protein called VWF, or von Willebrand Factor. VWF plays two roles in the clotting cascade. First and foremost, it protects Factor VIII from degradation in the liver. In the absence of VWF, the plasma half-life of Factor VIII drops from 8 hours to around 2 hours, in essence creating a Factor VIII deficiency. Second, VWF is exposed during vascular injury when endothelial cells are sheared off blood vessel walls. It helps anchor platelets at the site of damage, speeding up clot formation. Consequently, a deficiency of VWF results in an abnormally long bleeding time. In some cases, ddAVP (arginine vasopressin) treatment reduces bleeding episodes in patients with Von Willebrand disease; in other cases, Factor VIII injections may be helpful.

Several genetic disorders result in a hypercoagulable state; however, unlike the above-mentioned diseases, most people with these disorders are heterozygotes whose symptoms exhibit great variability.

The best characterized of these disorders include deficiencies of Antithrombin III, Protein C, Protein S, and Factor V Leiden mutation. The treatment for these disorders is heparin injections in the acute setting.

Long-term treatment consists of anticoagulation with warfarin for patients with Protein C, S, or Factor V Leiden deficiency. People who are homozygous for Protein C deficiency require periodic blood plasma transfusions rather than oral warfarin therapy.

The Role of Vitamin K

Vitamin K is a fat-soluble compound found in broccoli, green leafy vegetables like kale or Romaine lettuce, and animal liver. This vitamin was designated "K," stemming from the German word for coagulation, or blood clotting. Although intestinal bacteria produce some Vitamin K, the majority is obtained from dietary sources. The human body requires this vitamin to produce functional blood clotting proteins, particularly Factor II (prothrombin) and Factors VII, IX, and X.

Each of these proteins is a serine protease, an enzyme that cleaves certain peptide bonds in its target protein(s). As it turns out, these particular clotting factors must bind calcium in order to become fully activated. The ability to bind calcium, in turn, requires these proteins to be modified at certain glutamate residues, known as gamma-carboxylation. The amino acid glutamate normally contains a single negatively charged COO-, or carboxylic acid group in its side chain. An additional COO- group allows glutamate to bind positively charged calcium ions much more effectively. Vitamin K is an essential cofactor in the enzymatic reaction producing gamma carboxyglutamate.

The anticoagulant or blood thinner known as warfarin (or Coumadin) is a structural analog of Vitamin K. Its mechanism of action is to act as a competitive antagonist, in effect decreasing gamma-carboxylation of Factors II, VII, IX, and X by posing as Vitamin K. A therapeutic dose of warfarin prolongs blood coagulation time (a parameter called PT or prothrombin time) by a factor of 2 to 3. High doses of warfarin are dangerous because they raise the risk of spontaneous internal bleeding. As such, anyone on warfarin for an extended period of time is advised to be seen every one to two weeks at an AC or anticoagulation clinic.

Medically, injections of Vitamin K are indicated for all newborn babies to help their livers get up to speed in producing functional clotting factors. Alcoholics, malnourished people, and those suffering from liver disease are also given supplements of Vitamin K, usually by mouth. In these cases, Vitamin K helps prevent internal hemorrhages. In the case of people requiring TPN (Total Parenteral Nutrition), Vitamin K is combined with the other fat-soluble vitamins A, D, and E in a lipid formulation. Luckily, a person's liver stores enough Vitamin K to last a few months, making a deficiency of this vitamin less common than deficiencies of water-soluble vitamins (the B complex and Vitamin C).

Essay 52: White Blood Cell Surface Receptors in Innate Immunity

Neutrophils, monocytes, eosinophils, and basophils form the backbone of the body's innate immune system. Although these white blood cells differ in appearance, abundance, and longevity, they share some key features in common. First, all of these cells arise from non-lymphocyte precursors termed myeloid cells. Second, upon encountering foreign antigens, each cell type responds in a highly predictable, invariant manner; this response is the hallmark of innate immunity. This article will explore some of the cell surface receptors involved in the innate immune response, particularly emphasizing foreign antigen recognition, chemotaxis, and cell migration from the bloodstream into the surrounding tissues.

Innate immune cells become activated in three main ways: 1) they encounter foreign antigens directly, 2) they respond to signals called cytokines produced by other white blood cells, or 3) they bind to adhesion molecules expressed by vascular endothelial cells in regions of infection or inflammation.

In the first scenario, white blood cells express one or more classes of surface proteins called toll-like receptors (TLRs) that bind to bacterial and viral antigens. For example, monocytes express TLR-4, which binds to a bacterial membrane component called lipopolysaccharide (LPS). The cell instantly recognizes LPS as a foreign antigen and sounds the alarm, so to speak, by attracting other white blood cells to the site of infection.

In the second scenario, white blood cells follow a trail of soluble signals called cytokines that lead them to the site of infection. The ability to migrate in response to chemical signals is called chemotaxis. Some cytokines attract specific white blood cells; in contrast, others induce chemotaxis in multiple cell types. Leukotriene B4 and Interleukin-8 selectively target neutrophils, while a protein called RANTES attracts basophils and eosinophils.

Although the details of cytokine signaling are incompletely understood, it appears that many cytokine receptors are members of the JAK-STAT protein family.

The term JAK-STAT is an acronym for Janus Kinase / Signal Transduction and Activator of Transcription. Basically, when a cytokine signal binds to the JAK membrane receptor, its cytosolic domains phosphorylate the nearby STAT proteins, which proceed to pair up, travel to the cell nucleus, and turn on gene transcription.

In the final scenario, white blood cells exit the circulation and enter the extravascular tissues in a three-step process: First, the white blood cells roll along the endothelial lining of veins and capillaries until they come to a halt. Next, they squeeze themselves between the endothelial cells (diapedesis). Finally, the white blood cells migrate through the extracellular matrix to combat microbial invaders in the tissues.

This process, known as endothelial transmigration or extravasation, is governed by multiple sets of cell surface receptors. The best-characterized interaction involves proteins called integrins located on the white blood cell surface that bind to endothelial cell proteins called CAMs (cell adhesion molecules).

Meanwhile, a white blood cell receptor called the sialyl Lewis X antigen binds to a group of endothelial cell glycoproteins called selectins. E-selectin is found mostly on vascular endothelial cells. P-selectin was isolated from placental tissue, and L-selectin is most abundant in the spleen, lymph nodes, and other lymphocyte-rich tissues.

All in all, the innate immune system relies on a multitude of surface receptors to effectively serve as the body's first line of cellular defense.

Medicine & Disease

Essay 53: Alpha-1 Antitrypsin Deficiency

Alpha-1 antitrypsin deficiency (AATD) is an autosomal recessive genetic condition thought to affect an estimated 200,000 people in the United States today. Most people with AATD are of Northern European or Iberian ancestry. Men and women are affected in equal numbers. AATD results from lacking a particular protein called alpha-1 antitrypsin, or AAT. Normally, AAT counteracts the digestive actions of trypsin, a protease enzyme found in the gastrointestinal and respiratory tracts.

Trypsin and similar enzymes aid in the digestion of dietary proteins and may combat bacterial overgrowth in the gut and upper airways. Left unchecked, however, trypsin can damage the alveoli in the lungs as well as liver hepatocytes and epithelial cells lining the biliary tree. The end results of this enzymatic assault are emphysema and, in some cases, liver cirrhosis.

Symptoms

Unlike many genetic diseases, AATD rarely appears in childhood; rather, the vast majority of cases are diagnosed in early adulthood. Most people with AATD come to medical attention after developing symptoms of emphysema, sometimes before age 30, even though they often have no history of smoking. Occasionally, AATD manifests as jaundice (yellowing of the skin and sclera of the eyes) or as elevated liver enzymes in a patient with no known history of alcoholism or viral hepatitis.

Although the reason(s) for the delayed onset of AATD is not completely understood, one explanation is that the lungs have enough reserve alveoli to mask the problem for the first 20 to 30 years of a person's life. After enough damage has accumulated, however, the symptoms of emphysema show up as wheezing, shortness of breath, a dry cough, and vulnerability to infections like bronchitis and pneumonia. Unlike the lungs, the liver can regenerate to some extent. In some cases of AATD, the patient's liver compensates for the damage inflicted by unopposed trypsin activity. In more severe cases, or in patients who drink alcohol regularly, liver cells are destroyed faster than they can be replaced, resulting in jaundice and, eventually, cirrhosis.

Diagnosis

AATD is diagnosed by a blood test measuring serum levels of alpha-1 antitrypsin. In patients with borderline levels of AAT, PCR (polymerase chain reaction) can be performed to compare the DNA sequence of the patient's SERPINA-1 (Serpin Protease Inhibitor Alpha-1) gene to the wild-type sequence. Multiple variants of the SERPINA-1 gene exist, accounting for the broad spectrum of disease presentation in AATD patients.

Available Treatments

Currently, the main treatment available for AATD is injections of alpha-1 antitrypsin to compensate for the deficiency of this protein. With this treatment and respiratory therapy, non-smokers with mild or moderate AATD have a generally good prognosis, with a median survival of age 56.

Liver transplantation is a treatment of last resort in AATD patients with end-stage hepatic failure. Although lung transplants have been performed in AATD patients, lung allografts tend to be rejected despite therapy with steroids, cyclosporine, Tacrolimus, and other immunosuppressant drugs.

Gene replacement therapy may be a viable option in the near future. This may be accomplished by stem cell transplantation or possibly by using genetically engineered hepatitis viruses as vectors to carry functional AAT genes into the patient's liver cells.

Essay 54: Alcoholism

Alcohol abuse is estimated to affect 17.6 million people in the United States alone. Besides the harmful impact alcohol addiction has on families and society as a whole, the disease exacts a terrible toll on the physical health of the individual caught in its grip.

The most common health problem seen in heavy drinkers and chronic alcoholics is liver disease. An estimated 25% or more of alcoholics develop fatty liver followed by alcoholic hepatitis and ultimately by cirrhosis, an irreversible scarring of the liver parenchyma. Severe cases of cirrhosis can lead to fulminant hepatic failure, in which the liver's metabolic functions essentially collapse. Without a liver transplant, this severe form of liver failure results in hepatic encephalopathy, which leads to coma and death.

Cirrhosis is also a setup for portal hypertension, a vascular disorder marked by esophageal varices, which are dilated veins in the esophagus prone to bleeding; caput medusa or spider-shaped blood vessels on the surface of the abdomen; hemorrhoids; and, in some cases, a condition called ascites, the leakage of vascular fluid into the abdominal cavity due to deficient albumin production by the liver. Finally, cirrhosis predisposes a person to develop hepatocellular carcinoma (HCC), better known as liver cancer. With rare exceptions, HCC is fatal within 6 months or so of diagnosis.

A second gastrointestinal disorder relatively common among alcoholics is pancreatitis, especially the chronic form. The link between alcoholism and pancreatitis is not as well understood as liver disease. Chronic pancreatitis is a known risk factor for pancreatic cancer, one of the deadliest known malignancies. The 5-year survival rate is less than 1%.

Yet another organ system harmed by chronic alcohol abuse is the circulatory system. For obscure reasons, alcoholics are at higher risk for developing cardiomyopathy (CMO), a condition in which the heart muscle becomes baggy and dilated. CMO is a setup for arrhythmias and strokes. Even with treatment, CMO often progresses to congestive heart failure (CHF), with a 5-year survival rate of less than 50%.

Over time, alcoholics tend to become malnourished and suffer from poor dental hygiene. Both of these conditions predispose to various infections, especially gingivitis and dental abscesses, which raise one's risk of infective endocarditis.

The brain is not spared from the ravages of alcoholism. A significant number of chronic alcoholics develop a condition called Wernicke-Korsakoff syndrome. This is a neurological disorder characterized by nystagmus (involuntary eye movements), ataxic gait (walking in an uncoordinated manner), and psychiatric disturbances, especially confabulation, in which they invent plausible stories to fill in gaps in memory.

Wernicke-Korsakoff syndrome is believed to result from thiamine deficiency with subsequent damage to the cerebellum and a hypothalamus region called the mammillary bodies.

Thiamine deficiency can be corrected; the brain damage, however, is irreversible. In a final twist of irony, abrupt withdrawal from alcohol may lead to a potentially fatal disorder called delirium tremens (DTs). Mostly seen in chronic alcoholics, DTs may occur up to 10 days after the last drink and are marked by seizures, coma, and sometimes death. For this reason, alcohol withdrawal is managed with anticonvulsants, particularly the drug chlordiazepoxide (Librium).

Essay 55: Aging and the Brain Alzheimer's Disease

In healthy human aging, the signs of wear and tear on the brain are relatively nonspecific. Elderly people (even those not affected by Alzheimer's disease) tend to exhibit some degree of brain tissue atrophy with advancing age. In the case of brain tissue, atrophy is characterized by a decline in the number of functional neurons and associated glial cells.

The causes underlying this process are still a subject of debate but are probably related to the systemic vascular pathology seen in other chronic disorders, especially hypertension, and diabetes. The theory goes that as blood vessels age, the tissues they supply receive less oxygen and nutrients and may accumulate more metabolic waste products than younger people.

Beyond the wear and tear seen in normal aging, the major degenerative disease associated with advanced age is Alzheimer's disease (AD). This disease is characterized by abundant deposits of beta-amyloid plaque protein and neurofibrillary tangles in the brains of people examined at autopsy. In spite of major advances in neuroimaging, AD remains a clinical diagnosis that can only be confirmed by an autopsy postmortem.

Although AD was first described by Alois Alzheimer in 1906, the underlying cause remains unknown in most cases. A small percentage of AD cases display a hereditary pattern. People with familial AD may develop symptoms before age 50. The molecular culprit in these cases is believed to be an abnormal variant (or abnormal expression) of the Apo E gene located on chromosome 19. The Apo E gene encodes an apolipoprotein; this class of protein is involved in triglyceride metabolism in CNS neurons. Abnormal levels of Apo E somehow result in neuronal cell death over the course of years, but the details remain a mystery.

However, Most AD cases occur sporadically and are not associated with Apo E abnormalities. Three main camps of researchers continue to debate the cause(s) of AD. Some focus on the beta-amyloid plaques, which appear to be more prevalent in the brains of AD patients compared to their same-age counterparts. People born with Down's Syndrome (a trisomy of chromosome 21) tend to develop brain pathology marked by abundant beta-amyloid plaques if they live to age 50. A straightforward explanation is that the APP gene, which encodes Amyloid Precursor Protein, is located on Chromosome 21.

An extra copy of the gene probably accounts for the premature plaque deposits in the brains of people with Down's syndrome; however, it remains unclear how APP expression becomes aberrant in people with 2 copies of the gene.

Other researchers focus on the neurofibrillary tangles, which are believed to arise from aggregates of excessively phosphorylated tau protein, a microtubule-associated protein whose normal function remains uncertain. A third group believes the culprit is a protein called alpha-synuclein, which may also play a role in Parkinson's disease. Yet others think the emphasis on abnormal protein deposits is misguided.

Although amyloid plaques and NFTs may be a proximate cause of AD, the ultimate cause of the disease probably involves aberrant gene expression in neurons and glial cells. What ultimately triggers deranged gene expression in the CNS is, of course, anyone's guess. Recent theories invoke a host of factors ranging from chronically elevated glucose levels to free radical exposure to subclinical CNS hypoxia.

A professor of mine once quipped that the three camps could be labeled the BAP-tists (for whom the enemy, β amyloid plaque, is always found outside the cell), the Tau-ists (who focus on a lack of inner harmony), and the Syn-ers (convinced the enemy lurks within the cell nucleus).

He was an avowed agnostic (getting together is unimportant), by which he expressed his belief that abnormal protein aggregates are a late marker of Alzheimer's disease, not the causative factor. One or more of these proteins may yet turn out to be the major culprit in AD; nevertheless, the debate will continue for the foreseeable future.

Essay 56: Common Causes of Anemia

Anemia refers to an absolute reduction in the number of functional red blood cells in a person's bloodstream. It is a medical maxim that anemia is not a primary disease; instead, it is a marker of an underlying disease. In practice, most cases of anemia fall into the following categories: blood loss exceeding red blood cell (RBC) production capacity; a deficiency in one or more nutrients essential for RBC production; a bone marrow disorder; a genetic condition; and last, but not least, anemia secondary to a chronic disease state.

Blood Loss

The source of an acute bleed may be obvious, as during a traumatic incident like a motor vehicle accident. Other times, internal bleeding may be occult, as in patients with undiagnosed colon cancer or certain autoimmune disorders such as lupus or other forms of vasculitis (inflammation of the blood vessels). The most reliable treatment for acute blood loss is a blood transfusion.

Nutrient Deficiencies

One of the most common, but correctable, causes of anemia is iron deficiency. A shortage of iron leads to decreased hemoglobin production, in turn resulting in small, pale red blood cells. In medical jargon, this is termed hypochromic, microcytic anemia. People most at risk for iron deficiency anemia include women of reproductive age (due to blood loss with each menstrual period), strict vegetarians, and malnourished children, because they are still growing.

In general, men are at lower risk for iron deficiency because, unlike menstruating women, they do not lose significant amounts of blood at monthly intervals. Aside from blood loss, no other physiological pathway is known for iron disposal in humans. Iron can be replaced orally with tablets or by parenteral injections in patients who cannot tolerate oral iron supplements.

Two other nutrient deficiencies are often underlying causes of anemia: Vitamin B12 and folic acid. In contrast to iron deficiency, a shortage of B12 or folic acid leads to megaloblastic anemia. In both cases, a blood smear reveals abnormally large, hyperpigmented red blood cells and neutrophils with hypersegmented nuclei. Neurologic symptoms occur in B12 deficiency because this vitamin is required for the proper metabolism of fatty acids, a prominent component of the myelin sheath that insulates neuronal axons.

People at risk for low levels of these vitamins include patients with pernicious anemia (in which Vitamin B12 absorption is impaired); patients with extensive intestinal resections or inflammatory bowel disease; as well as alcoholics and chronically malnourished people who lack green, leafy vegetables (a major source of folic acid) in their diets. An absolute deficiency of B12 is relatively uncommon but can occur in strict vegetarians who exclude dairy products and eggs from their diets. Folic acid can be supplemented orally. As an aside, pregnant women are routinely prescribed folic acid tablets to minimize the risk of fetal neural tube defects. Vitamin B12, on the other hand, is usually replaced by monthly intramuscular injections in patients with pernicious anemia and intestinal disorders.

Bone Marrow Disorders

Leukemia, multiple myeloma, and certain myelodysplastic syndromes represent the major primary bone marrow disorders manifesting as anemia. Secondary bone marrow failure may be due to aplastic anemia, a life threatening condition that sometimes occurs in the wake of cancer chemotherapy or exposure to various other toxins.

Depending upon the cause, bone marrow failure may be self limited, but sometimes (especially in cases of leukemia and aplastic anemia), the only definitive treatment remains a bone marrow or stem cell transplant.

Genetic Diseases

This category encompasses disorders of hemoglobin synthesis, including sickle cell anemia, alpha thalassemia, and beta thalassemia, as well as rarer conditions such as Fanconi's anemia, a severe pediatric disease marked by pancytopenia, the defective production of all blood cell types. Treatments for sickle cell anemia include hydroxyurea along with vaccinations against encapsulated bacteria, in particular streptococcal pneumonia, H. influenzae, and meningococcus. Patients with full blown beta thalassemia require monthly blood transfusions to survive. Children with Fanconi's anemia are treated with blood transfusions and either bone marrow or hematopoietic stem cell transplants.

Anemia of chronic disease

The most common context of anemia of chronic disease is a patient with chronic kidney disease. Although the details are not fully understood, most scientists believe the problem stems from decreased production of the hormone erythropoietin, synthesized by Lacis cells in the kidneys. In healthy people, erythropoietin boosts RBC production in the bone marrow in response to a drop in blood oxygen levels. In chronic renal failure, however, the kidneys synthesize little or no erythropoietin, putting these patients at high risk of developing anemia.

Patients with various forms of cancer may also develop anemia as part of the wasting syndrome (cachexia) that can occur in cases of advanced malignancy. Although the cause(s) remain obscure, it is possible that in metastatic cancer, the tumor cells act like energy parasites, diverting oxygen and nutrients away from the bone marrow, making anemia practically inevitable.

Essay 57: Arthritis Explained

The definition of arthritis is inflammation of a joint or joints. It is useful to distinguish the different categories of arthritis as well as the treatments that exist.

1) Osteoarthritis (OA) - this is the run of the mill form of arthritis that can be thought of as resulting from the wear and tear of old age. Cartilage wears away on the bony surfaces of weight bearing joints resulting in the all too familiar symptoms: pain and stiffness, esp. with prolonged physical activity. Knees, hips, and the vertebral column are the main weight bearing joints of the body affected by osteoarthritis. Over 90% of people will have symptomatic OA by age 75.

Treatments for OA include analgesics, especially NSAID painkillers like aspirin, injections of cartilage components like glucosamine and chondroitin sulfate, and surgical intervention such as knee and hip replacements.

2) Rheumatoid arthritis (RA) - although the specific cause is unknown, the consensus is that RA is an autoimmune disease. Approximately 1% of the general population suffers from RA; the incidence is believed to be highest in women ages 20-40. The hallmark of RA is pain and stiffness in the smaller joints (especially hands, wrists, and ankles) that is worse in the morning and subsides over the span of an hour or so. For no clear reason, the DIP joints - distal interphalangeal joints closest to the fingertips - are seldom if ever affected in RA.

The treatment strategy in RA attempts to halt the immune onslaught and prevent irreversible joint erosions. Steroids like cortisone are used to control acute flare ups of RA. Older treatments for refractory RA included immune suppressant drugs like methotrexate and gold salts, but due to their dangerous side effects (bone marrow suppression, kidney damage) they are used as drugs of last resort today.

More recently, mononclonal antibody drugs like Humira (Adalimumab), Remicade (Infliximab) and Etanercept, which binds to and neutralizes the inflammatory cytokines TNF-α (Tumor Necrosis Factor alpha) and Interleukin-1, have come into widespread use. Monoclonal antibodies have become the standard treatment for juvenile rheumatoid arthritis (JRA) as well.

3) Psoriatic arthritis - as the name suggests, this form of arthritis accompanies the skin disease psoriasis. Arthritis may also occur in the context of other autoimmune diseases including lupus and Crohn's disease.

Treatment is mainly aimed at pain relief. If the underlying disease is brought under control, the arthritis often subsides.

4) Infectious arthritis - the main disease that falls into this category is an untreated strep infection resulting in rheumatic fever. Another cause of infectious arthritis is untreated gonorrhea (Reiter's syndrome) that can affect one or more joints. Treatment consists of a course of antibiotics (e.g. Azithromycin and Ceftriaxone) to eradicate the bacterial infection.

Essay 58: Autopsy Basics

In decades past, autopsies were routinely performed on people who died in U.S. hospitals. Today, autopsies are largely restricted to forensics cases or are performed at the behest of the next of kin. As will be discussed, the general procedure of an autopsy is largely the same regardless of the reason for its performance.

The first step of an autopsy is to positively identify the deceased and note the time of death. This information is readily available for virtually all hospital patients. Unidentified bodies recovered outside of a hospital are stored separately in the morgue or in a decomposition room. Forensics teams usually handle these cases, which are considered homicides until proven otherwise.

The next step is to visually inspect the patient's body for signs of blunt force or penetrating trauma, burns, needle marks, and other skin lesions. The presence of any tattoos or large scars is also noted. If any trauma or gross lesions are present, especially on the head or neck, these areas are photographed. The oral cavity is visualized and the patient's dentition is noted. With the advent of PCR (Polymerase Chain Reaction) and DNA identification, there is seldom a need to take dental impressions, and so the autopsy continues with a thoracotomy.

Chest Cavity

A bone saw is used to cut through the ribs and expose the thoracic cavity. The pathologist purposely avoids bisecting the sternum, mainly to avoid incidental damage to the heart and surrounding structures. The lungs are examined for the presence of primary or metastatic cancer as well as signs of other pulmonary diseases, for example tuberculosis. Collapsed lung tissue tends to be friable, disintegrating readily unless soaked in formaldehyde. The diaphragm muscle is examined along with the contents of the mediastinum, namely the esophagus and great vessels, i.e. the aorta, pulmonary artery, superior vena cava, and inferior vena cava.

After severing the great vessels, the heart is removed from the pericardial sac and weighed. Even in the absence of overt cardiac disease, other factors must be considered when examining the heart, for example, the patient's ideal vs. actual body weight, comorbid illnesses, and general state of health at the time of death. An abnormally enlarged heart, known as cardiomegaly, is encountered frequently on postmortem examination. Major disease states resulting in cardiomegaly include chronic hypertension, congestive heart failure (CHF), valvular disease (especially aortic stenosis), rheumatic fever, and severe COPD, resulting in right sided heart enlargement, also called cor pulmonale. Any evidence of past myocardial infarctions (heart attacks) or open heart surgery is noted as well. Samples of cardiac muscle, and occasionally pericardial tissue, are sent to pathology.

Abdominal cavity

The examination of the abdominal cavity and its contents is often the highlight of the autopsy. After incising the membranous covering of the abdomen called the omentum, the organs of the gastrointestinal tract are examined and weighed, including the stomach, liver, gallbladder, and pancreas. Cirrhotic livers appear yellow and abnormally small. At the other extreme, patients with CHF may exhibit hepatomegaly, or an enlarged liver, due to congestion of the hepatic sinusoids with blood. The spleen is then removed and weighed, followed by both kidneys. At this point, the pathologist usually examines the small and large bowels, noting any structural anomalies, tumors, or the presence of foreign bodies in the intestinal lumen. Once the intestines have been removed, the abdominal aorta, now clearly visible, can be inspected for aneurysms or evidence of dissection.

Pelvis and Extremities

Unless there are signs of trauma or pathology involving the urinary bladder or reproductive tract, this part of the body receives little attention. The same holds true of the upper and lower extremities.

Postmortem blood work

In cases of a suspected drug overdose, an examination of gastric contents often reveals the lethal agents. Blood samples are sent for a toxicology screen, which detects levels of blood alcohol in addition to metabolites of opiates, cocaine, amphetamines, barbiturates, hallucinogens, marijuana, and over the counter drugs like aspirin and acetaminophen. The levels of certain prescription drugs, e.g. digoxin, lithium or benzodiazepines (tranquilizers) can also be measured.

In a full autopsy, the skull is sawed open and the brain (or its remnants) is removed. The brain is preserved for 10-14 days in formaldehyde. After preservation, sections of brain are cut and frozen for histological examination. Brains obtained from autopsies are also donated to medical schools for dissections conducted in gross anatomy or neuroanatomy courses.

Post-autopsy procedures

After the autopsy is completed, the patient's internal organs are returned to their appropriate locations. The thoracic and abdominal incisions are sutured, and the body is returned to the hospital morgue. Except for cases of suspected homicide, the patient's body is released to the next of kin for burial or cremation. All states in the U.S. mandate the donation of unclaimed cadavers to medical schools for anatomic dissections.

Cadavers unfit for scientific use are cremated or buried depending on the law of that particular state. In most states, the time limit applied to unclaimed bodies is 72 hours. One exception is Oregon, which waits 10 days before disposing of an unclaimed body. The federal government seldom interferes with state laws regarding unclaimed bodies unless there is reason to believe the deceased individual was a U.S. military veteran.

Essay 59: Cardiomegaly

The term cardiomegaly refers to an abnormally enlarged heart. In most cases, an enlarged heart results from a chronic problem as opposed to an acute etiology. Although many disease states can lead to cardiomegaly, it is useful to categorize them in terms of their effect on the cardiac muscle layer or myocardium. These underlying pathological changes, termed hypertrophic, dilated, and restrictive cardiomyopathy, respectively, can all lead to heart enlargement, and each will be explored in this article.

In cardiac hypertrophy, an excessive pressure load on the heart's left ventricle leads to an overgrowth of the myocardium. The two leading culprits in this scenario are uncontrolled hypertension and aortic valve stenosis. In both cases, the left ventricle must generate extra contractile force to eject the same volume of blood into the aorta. New cardiac muscle fibers are laid down parallel in response to pressure overload. The heart gradually enlarges until the muscle layer outgrows its own blood supply, at which point heart failure ensues. However, it is important to note that this form of cardiomegaly is partially reversible if a person's blood pressure is brought under control or after surgical replacement of the diseased aortic valve.

In contrast to hypertrophy, dilated cardiomyopathy results from volume overload in the left ventricle. Over time, as the left ventricle is forced to accommodate an abnormally large volume of fluid, the muscle layer becomes thin, baggy, and distended, owing to the serial deposition of new muscle fibers. Far and away, the main risk factor predisposing a person to dilatation of the left ventricle is a prior heart attack (myocardial infarction).

Valvular diseases are also a major culprit, in particular mitral and aortic valve regurgitation. Entire volumes have been written about heart valve disease; the important causes include congenital valve defects, rheumatic fever, any form of endocarditis, and calcification of the valve leaflets due to advanced age. Certain viral infections, especially the Coxsackie virus, are also associated with dilatation of the left ventricle. Chronic alcoholism is yet another risk factor for dilated cardiomegaly.

Cardiomegaly due to restrictive heart disease is far less common compared to the above two scenarios. The main culprits in this category include pericarditis and amyloidosis. Pericarditis refers to an inflammation of the membranous sac surrounding the heart.

Causes of pericarditis include a host of bacterial and viral infections, autoimmune diseases, myocardial infarction, and metastatic cancer.

Amyloidosis is a strange immunological disorder in which the liver produces an array of proteins that white blood cells fail to break down completely. These proteins proceed to form insoluble aggregates in many organs, including the liver, kidneys, spleen, and heart. Some cases of amyloidosis occur in conjunction with rheumatoid arthritis, inflammatory bowel disease, and multiple myeloma. Except for a heart transplant, no curative treatment exists for cardiac amyloidosis.

A few disease states do not fall neatly into any single category but still deserve mention. Congenital heart defects, especially the Tetralogy of Fallot, are often marked by cardiac enlargement. Several mitochondrial diseases predispose to cardiac enlargement and heart failure in young adulthood. Finally, Pompe's disease, a rare lysosomal disorder marked by incomplete sugar digestion, tends to result in cardiomegaly and death in early childhood or adolescence.

Isolated right-sided heart enlargement also merits an explanation. Most cases result from cor pulmonale, a condition most often seen in patients with severe COPD (chronic obstructive pulmonary disease) or other lung diseases such as silicosis, asbestos exposure, or pulmonary fibrosis. Nonetheless, most cases of right-sided heart enlargement occur in the context of left-sided heart failure.

Essay 60: Chromosomal Aneuploidy

The term aneuploidy refers to any abnormal karyotype or chromosomal endowment. Except for gametes and the XY sex chromosomes in males, human cells are normally diploid, meaning they contain two copies of the other 22 chromosomes (referred to as autosomes). In humans, some cases of aneuploidy involve monosomies, meaning the deletion of a chromosome during or shortly after fertilization. Trisomies (and rarely tetrasomies) also occur. Abnormal gene expression and chromosomal transport during mitotic cell division are thought to account for the anomalies observed in these syndromes; gene mutations themselves play, at most, a minor role.

Monosomies

The loss of all or part of a chromosome results in distinct developmental abnormalities. In Turner Syndrome, the deletion of an X chromosome (karyotype 45 XO) results in a female with short stature and a rudimentary or non-existent reproductive tract. Without estrogen replacement therapy, females with Turner syndrome seldom develop secondary sexual characteristics. Turner syndrome occurs in an estimated 1 in 2500 female births. In contrast to Turner's syndrome, no complete autosomal monosomy is compatible with life. A partial deletion of chromosome 5 results in a severe developmental disorder called Cri-du-chat (cry of the cat), named for the unusual-sounding cries of infants with this syndrome.

Autosomal Trisomies

The main trisomies compatible with life include Down syndrome (Trisomy 21), Patau syndrome (Trisomy 13), and Edwards syndrome (Trisomy 18).

Down syndrome is the most common autosomal trisomy in humans, occurring in approximately 1 in 800 births. Down syndrome is instantly recognizable; its features include short stature, a small head, close-set eyes, a lolling tongue, and mental retardation ranging from mild to severe. Around 50% of babies born with Down syndrome have cardiac defects. They are also at a higher risk of developing leukemia as well as early-onset dementia.

The reasons underlying the higher incidence of leukemia remain obscure; abnormal mitoses in leukocyte precursors probably play a role. As far as dementia goes, chromosome 21 is the site of the APP (Amyloid Precursor Protein) gene. The consensus among geneticists is that an extra copy of the APP gene results in dementia similar to Alzheimer's disease in most Down syndrome patients who live to age 40.

The other two autosomal trisomies produce far more severe abnormalities compared to Down syndrome. To put it in perspective, 50% or more of babies born with Down syndrome survive to young adulthood. In comparison, few infants with Patau or Edwards syndrome survive beyond the first year of life. Defective development of the heart, lungs, and kidneys is the major cause of mortality.

Pallister-Killian Syndrome

This extremely rare condition results from a partial tetrasomy of chromosome 12. For unclear reasons, the person's tissues exhibit mosaicism; some of the cells have a normal karyotype, whereas others contain an extra, aberrant copy of chromosome 12 called an isochromosome, in which the p arm of the chromosome is duplicated, resulting in four copies of 12p genes in each affected cell. People born with PKS have craniofacial malformations, moderate to profound mental retardation, and in many cases, diaphragmatic hernias, which impair the normal development of the heart and lungs.

Klinefelter Syndrome, XYY, and XXX: Sex Chromosome Trisomies

The most common polysomy involving the sex chromosomes is a disorder known as Klinefelter's disease. Klinefelter results from the non-disjunction of one or more X chromosomes during meiosis, leading to an abnormal karyotype (chromosome number or structure). The majority of non-disjunction events are believed to occur during oocyte formation; however, polyploidy may also occur during spermatogenesis but tends to result in monosomies, especially Turner's Syndrome (karyotype 45 XO).

Over 80% of individuals with Klinefelter's have a chromosomal karyotype of 47 XXY, as opposed to the normal male karyotype of 46 XY. Less often, polysomies XXXY and XXYY are observed in association with Klinefelter's. All polysomies in which a Y chromosome is present result in the development of male reproductive tract structures exclusively. Scientists think that even a single copy of certain Y chromosome genes, such as AMH (Anti-Mullerian Hormone), is sufficient to inhibit the formation of female gonads.

Most of the time, Klinefelter produces few distinct anatomical symptoms before puberty. By adolescence, however, males with Klinefelter tend to be over the 90th percentile in height and begin to develop several feminine features, including a broad pelvic diameter, lack of facial hair, and gynecomastia (growth of breast tissue). In most cases, males with Klinefelter's are infertile owing to testicular atrophy, which produces few or no sperm.

In contrast to Klinefelter's, the karyotype 47 XYY (an extra Y chromosome) is not associated with any specific anatomic abnormalities. This polysomy arises exclusively during spermatogenesis since oocytes carry only X chromosomes.

This condition may occur in as many as 1 in 1000 live male births; however, since 47 XYY males appear developmentally normal and can father children, most cases probably go undetected. The 47 XYY karyotype is associated with a higher incidence of learning disabilities and autism spectrum disorders.

The polysomy XXX results in an anatomically normal female. The 47 XXX karyotype may occur in as many as 1 in 1000 live female births; however, as with 47 XYY, most cases go undetected. A small percentage of women with 47 XXX experience premature ovarian failure, marked by irregular menses and early menopause (prior to age 40).

As in autosomal trisomies such as Down's Syndrome, advanced maternal age is a risk factor for Klinefelter's and 47 XXX. Advanced paternal age may also increase the risk of chromosomal non-disjunction; however, in the case of polysomy 47 XYY, most instances are believed to arise as sporadic germ-line events independent of a man's age.

Essay 61: Colon Cancer Risk Factors

Although colon cancer claims the lives of more than 50,000 Americans each year, the risk factors behind most cases of this disease are not as well characterized as the relationship between smoking and lung cancer. Most cases of colon cancer occur in people over the age of 50; however, hereditary forms of the disease can strike at a much younger age. Nonetheless, colon cancer prior to puberty is extremely rare.

Broadly speaking, the risk factors for colon cancer (and all other cancers) can be classified as genetic or environmental. According to medical experts, as many as 25% of colon cancers arise as part of a hereditary syndrome. The most common form of hereditary colon cancer is actually the less well-studied non-polypoid variety known as HNPCC (Hereditary Non-Polypoid Colon Cancer), also called Lynch Syndrome. The inheritance pattern is autosomal dominant, meaning anyone with this disease has a 50% chance of passing it on to their children. Women with HNPCC are at higher risk for developing endometrial and ovarian tumors.

The other hereditary colon cancer syndromes are also autosomal dominant. Most are characterized by the proliferation of numerous intestinal polyps, one or more of which becomes malignant. This is especially true of FAP (Familial Adenomatous Polyposis) disease, in which a mutation or deletion of the APC gene makes colon cancer all but inevitable by age 30. The standard treatment for a patient with FAP is to undergo a prophylactic colectomy.

Other diseases in this category (Turcot's syndrome and Gardner's syndrome) are characterized by polypoid colon cancers in addition to extraintestinal cancers involving the brain and musculoskeletal system, respectively. Other hereditary malignancy syndromes like Li-Fraumeni syndrome, which results from mutations in the p53 gene, also raise the risk of intestinal cancer.

The other 75% of colon cancers seem to be caused by environmental factors. These risk factors include smoking, a diet high in animal fats and low in fiber, and obesity. Epidemiologists have long noted a lower incidence of colon cancer in people who subsist on a high-fiber diet, e.g., a traditional Japanese diet, compared to their Western counterparts. A lack of dietary fiber leads to slower transit times for intestinal contents.

Supposedly, this increases the length of time for dietary carcinogens and free radicals to wreak havoc on the colonic epithelial cells, triggering the genetic mutations that lead to the formation of polyps and, ultimately, cancer.

The major weakness of the above theory is that the cells lining the human intestine are replaced once every 7-10 days. Nevertheless, there is no doubt that people who consume a low-fat, high-fiber diet suffer fewer cases of colon cancer. Beyond dietary fiber, there is the possibility that some of these populations consume foods rich in antioxidants, for example, green tea, which is popular in China and Japan. All in all, avoiding obesity, tobacco smoke, and excessive dietary fat intake are three ways of minimizing one's risk of developing colon cancer.

Essay 62: Common Causes of Anemia

Anemia refers to an absolute reduction in the number of functional red blood cells in a person's bloodstream. It is a medical maxim that anemia is not a primary disease but a marker of an underlying disease. In practice, most cases of anemia fall into the following categories: blood loss exceeding red blood cell (RBC) production capacity; a deficiency in one or more nutrients essential for RBC production; a bone marrow disorder; a genetic condition; and last but not least, anemia secondary to a chronic disease state.

Blood Loss

The source of an acute bleed may be obvious, as during a traumatic incident like a motor vehicle accident. Other times, internal bleeding may be occult, as in patients with undiagnosed colon cancer or certain autoimmune disorders such as lupus or other forms of vasculitis (inflammation of the blood vessels). The most reliable treatment for acute blood loss is a blood transfusion.

Nutrient Deficiencies

One of the most common but correctable causes of anemia is iron deficiency. A shortage of iron leads to decreased hemoglobin production, in turn resulting in small, pale red blood cells. In medical jargon, this is termed hypochromic, microcytic anemia. People most at risk for iron deficiency anemia include women of reproductive age (due to blood loss with each menstrual period), strict vegetarians, and malnourished children because they are still growing.

Generally, men are at lower risk for iron deficiency because, unlike menstruating women, they do not lose significant amounts of blood monthly. Aside from blood loss, no other physiological pathway is known for iron disposal in humans. Iron can be replaced orally with tablets or by parenteral injections in patients who cannot tolerate oral iron supplements.

Two other nutrient deficiencies are often underlying causes of anemia: Vitamin B12 and folic acid. In contrast to iron deficiency, a B12 or folic acid shortage leads to megaloblastic anemia. A blood smear reveals abnormally large, hyperpigmented red blood cells and neutrophils with hypersegmented nuclei in both cases. Neurologic symptoms occur in B12 deficiency because this vitamin is required to properly metabolize fatty acids, a prominent component of the myelin sheath that insulates neuronal axons.

People at risk for low levels of these vitamins include patients with pernicious anemia (in which Vitamin B12 absorption is impaired), patients with extensive intestinal resections or inflammatory bowel disease, as well as alcoholics and chronically malnourished people who lack green, leafy vegetables (a major source of folic acid) in their diets. An absolute deficiency of B12 is relatively uncommon but can occur in strict vegetarians who exclude dairy products and eggs from their diets. Folic acid can be supplemented orally. As an aside, pregnant women are routinely prescribed folic acid tablets to minimize the risk of fetal neural tube defects. Vitamin B12, on the other hand, is usually replaced by monthly intramuscular injections in patients with pernicious anemia and intestinal disorders.

Bone Marrow Disorders

Leukemia, multiple myeloma, and certain myelodysplastic syndromes represent the major primary bone marrow disorders manifesting as anemia. Secondary bone marrow failure may be due to aplastic anemia, a life-threatening condition that sometimes occurs in the wake of cancer chemotherapy or exposure to various other toxins. Depending upon the cause, bone marrow failure may be self-limited, but sometimes (especially in cases of leukemia and aplastic anemia), the only definitive treatment remains a bone marrow or stem cell transplant.

Genetic Diseases

This category encompasses disorders of hemoglobin synthesis, including sickle cell anemia, alpha thalassemia, and beta thalassemia, as well as rarer conditions such as Fanconi's anemia, a severe pediatric disease marked by pancytopenia, the defective production of all blood cell types. Treatments for sickle cell anemia include hydroxyurea and vaccinations against encapsulated bacteria, particularly streptococcal pneumonia, H. influenzae, and meningococcus. Patients with full-blown beta thalassemia require monthly blood transfusions to survive. Children with Fanconi's anemia are treated with blood transfusions and either bone marrow or hematopoietic stem cell transplants.

Anemia of chronic disease

The most common context of anemia of chronic disease is a patient with chronic kidney disease. Although the details are not fully understood, most scientists believe the problem stems from decreased production of the hormone erythropoietin, synthesized by Lacis cells in the kidneys. In healthy people, erythropoietin boosts RBC production in the bone marrow in response to a drop in blood oxygen levels. In chronic renal failure, however, the kidneys synthesize little or no erythropoietin, putting these patients at high risk of developing anemia.

Patients with various forms of cancer may also develop anemia as part of the wasting syndrome (cachexia) that can occur in cases of advanced malignancy. Although the cause(s) remain obscure, it is possible that in metastatic cancer, the tumor cells act like energy parasites, diverting oxygen and nutrients away from the bone marrow, making anemia practically inevitable.

Essay 63: Common Myths about Medical School Acceptance

Myths abound about the gauntlet that medical school applicants must survive just to become medical students. Hearing people describe the process sounds as harrowing as basic training in the Marine Corps. On the other hand, every legend contains some truth. So, let's cut through the layers of exaggeration and get to the kernels of truth.

Myth #1 - You need a 4.0 GPA in college to get into medical school.

Reality: Although getting straight A's can't hurt, it is unrealistic for medical schools to consider the tiny pool of applicants who maintained a 4.0 GPA throughout their undergraduate years. Also, in the eyes of the admissions committee, getting a B+ at a school with name recognition, e.g., Harvard or Johns Hopkins, is the equivalent of getting an A at virtually any other school. I can't substantiate that last claim, but it was the rumor from where I went to med school.

Myth #2 – Minorities have an easier time being accepted to medical school because of quotas and other unwritten affirmative action policies.

Reality: Without taking a long detour into the touchy territory of race relations and affirmative action, suffice it to say that minorities do NOT enjoy a significant advantage in the medical school selection process. Assuming that a particular school sets aside ten to twenty percent of its available spots in the incoming class for minorities, this translates into no more than forty openings, even at the medical schools with the largest class size (approximately 200 people). Remember that most medical schools accept just over 100 new students per year. Considering there are approximately 110 medical schools in the United States, this works out to something like 4,000 out of 16,000 openings awarded to minorities in any given year.

Second, and more importantly, once a person is accepted to medical school, it's sink or swim. Medical schools hold students to a higher academic achievement standard than most college students are accustomed to. If you can't measure up, you're gone. They don't care if you're white, black, green, or purple.

Myth #3 – The interviewers pay attention to the smallest details of the attire people wear to med school interviews and often use this as the deciding factor in acceptance or rejection. One person

I knew at Hopkins was convinced that any guy who wore a shirt with buttons on the ends of the collar would be rejected immediately.

Reality: Every med school applicant I've seen wears a conservative, dark-colored suit or dress. Like you've heard umpteen times, dress for success. You're good to go if you've bathed, shaved and brushed your hair. Just remember, looks alone will not get you in. After all, you're competing in a contest to go to medical school, not the Mr. / Miss Universe pageant.

Myth #4 – Taking the MCAT (Medical College Admission Test) more than once is suicide because medical schools will judge an applicant based on his/her lowest score.

Reality: Although I have no direct knowledge of how medical schools make their choices, I tend to believe their claim that no one factor outweighs all others in their decision to accept an applicant.

An awesome MCAT score would be of little value if one's college academic performance were dismal. Conversely, an applicant with a solid GPA and average MCAT score is almost certain to be invited to one or more interviews.

Medical school application committees tend to focus on the big picture. Hence, if a person takes a qualifying exam, gets an average score, retakes the exam, and does significantly better, normal people would interpret this to mean that the applicant is striving to improve his/her chances of being accepted. From my experience, the vast majority of the faculty at the Penn State College of Medicine consisted of normal people. Nobody's perfect except for this guy in the Radiology department who looked like a beluga whale.

As an aside, most people don't know that the MCAT has almost nothing to do with medical science. Its main purpose is to measure how well you can answer many multiple-choice questions in a limited amount of time. In short, it is a numerical measure of performance under stress. Just as SAT scores cease to matter the moment a person enters college, the MCAT has no predictive value of how well an individual will do once s/he starts medical school.

Myth #5 – Non-science majors have an advantage in the application process because they will stand out for having focused on the liberal arts during their undergraduate years.

Reality: I'm not one to bash the liberal arts – whatever this nebulous term encompasses – but the truth is, the more time you devote to undergraduate biology courses, the less you'll have to struggle during the first two years of medical school. Another disadvantage of being a non-science major in college is that you have to fulfill not only the requirements of that particular major but also the prerequisite courses that all medical schools want to see, including chemistry, organic chemistry, cell biology, physics, and the corresponding lab courses. Even worse, you probably will not get credit for a double major even though you have done an equivalent amount of work.

I advise taking courses in literature, sociology, and history largely as GPA boosters. Think of it this way. You'll be doing plenty of reading in medical school; all sociology is good for is stating the glaringly obvious. Chances are, you'll learn more history watching the History Channel than any history lecture.

I hope I've cleared up some common misconceptions about the med school application process. For those of you who do not make it into medical school the first time around, decide if medicine is what you really want, then brush yourself off and try again. For those of you who run the gauntlet and make it into medical school, congratulations, it's show time.

Essay 64: Crohn's Disease

Crohn's disease (CD) is considered to be a form of inflammatory bowel disease (IBD). Although the cause/s of CD remains unknown, the epidemiology of the disease is well described. CD is more common in Caucasian populations as well as in industrialized countries. The highest incidence of CD is believed to occur among Ashkenazi Jews (meaning Jews who trace their ancestry to Central and Eastern Europe). In the U.S., the prevalence of CD is estimated at 3 per 100,000 in the general population and 10 per 100,000 among Ashkenazi Jews. The onset of CD peaks between adolescence and age 35. CD is less common in the elderly and vanishingly rare in children. Men and women are affected in roughly equal numbers.

In addition to Jewish ancestry, established risk factors for CD include a family history of the disease and smoking. Although many scientists suspect a bacterial or viral culprit in CD, no convincing evidence exists to support an infectious etiology. Some scientists go to the opposite extreme and claim that the true culprit is an overly sanitary environment. According to this view, without worms and other parasites to fight on a regular basis, the human immune system becomes bored for lack of a challenge and then goes haywire at the slightest provocation. Proponents of this argument point out that in many developing countries, intestinal parasites are common, but autoimmune diseases like CD, ulcerative colitis, asthma, and most allergies are distinctly uncommon. The debate continues with no end in sight.

Pathology/Symptoms

The hallmark of CD is an autoimmune attack on any part of the gastrointestinal tract. The white blood cells carrying out the assault produce transmural inflammation, meaning all bowel wall layers are damaged. The symptoms of CD vary widely. In mild cases, patients report occasional diarrhea and abdominal pain. More severe cases are characterized by fever, weight loss, bowel obstruction, intestinal adhesions, fistula formation, or abnormal connections between loops of the bowel or other organs, including the urinary bladder and skin.

In addition to gastrointestinal problems, some patients with CD suffer from arthritis and skin lesions. These extraintestinal manifestations suggest that CD may actually be a systemic autoimmune disease. As with other systemic autoimmune diseases like sarcoidosis and lupus, aberrant activation of B and T lymphocytes seems to be a common feature.

Treatment Options

CD is often difficult to treat. Flare-ups generally respond well to systemic corticosteroids, e.g., hydrocortisone; however, chronic steroid treatment leads to major side effects, including obesity, diabetes, osteoporosis, and vulnerability to infections. The anti-inflammatory drug 5-ASA (an aspirin derivative) is sometimes used in CD but is more often used to treat ulcerative colitis.

Other immunosuppressant agents like methotrexate and cyclosporine are often helpful in severe cases of CD. As with steroids, these drugs have serious side effects when used on a long-term basis. Newer medications include monoclonal antibodies such as adalimumab (Humira). Surgery is usually not curative in the case of CD because the disease can recur unpredictably in any part of the GI tract.

Essay 65: Diabetes Mellitus – Scrape the Icing Off the Cupcake

Physicians categorize diabetes mellitus (DM) as Type 1 (formerly called insulin-dependent or juvenile-onset diabetes), Type 2 (formerly called non-insulin-dependent or adult-onset diabetes), and gestational diabetes. DM can also occur secondary to other endocrine disorders like Cushing's syndrome. Because the pathogenesis and risk factors differ for each form of DM, we shall consider each separately.

Type 1 DM accounts for approximately 10% of diabetes cases; most cases come to medical attention by adolescence. This form of diabetes results from destroying the insulin-producing beta cells found in the Islets of Langerhans that make up the endocrine pancreas. The immune system mediates the destruction; as such, Type I DM is generally considered to be an autoimmune disease. When 90% of the beta cells are destroyed, the person develops the classic symptoms of insulin deficiency: weight loss, fatigue, excessive thirst (polydipsia), excessive hunger (polyphagia), and excessive urine output (polyuria).

Although cases of Type 1 DM have been described since ancient times, the risk factors for this disease remain unclear. Some speculate that exposure to certain viruses during childhood provokes an autoimmune reaction, destroying the islet cells. Some scientists consider people with the HLA protein DQ3.1 at higher risk for Type 1 DM. However, neither theory seems to carry much weight, as no specific infectious agent has been linked to the disease.

Also, most people with the DQ3.1 haplotype never develop Type 1 DM. Regarding genetic predisposition or vulnerability, most cases of Type 1 DM occur sporadically. Although a positive family history should not be discounted altogether, no clear pattern of heredity seems to exist for this form of diabetes.

In contrast to Type 1 DM, Type 2 DM has clear-cut risk factors: a positive family history and obesity. Type 2 DM accounts for nearly 90% of cases of diabetes in the US and the industrialized world. It is believed to result from the person's liver, skeletal muscles, and fat tissue becoming progressively unresponsive to the actions of insulin. In medical jargon, this is known as end organ insensitivity.

Although Type 2 DM can often be controlled with diet, exercise, and oral hypoglycemic medications (metformin, glipizide), 25% or more of people with the disease will eventually need insulin injections to manage their blood sugar. Type 2 DM seldom strikes before age 40; clinicians have reported Type 2 DM cases in increasingly younger patients, especially in morbidly obese adolescents.

The third category, gestational diabetes, occurs in pregnant women. Although positive family history and obesity probably raise the risk of gestational DM, most women who develop this condition have no obvious risk factors. According to one theory, most pregnant women become insensitive to insulin to some degree, owing to the effects of a host of elevated hormones, including hCG (human chorionic gonadotropin), hCS (human chorionic somatotropin), estrogen, and other steroid hormones.

Most pregnant women are given an oral glucose tolerance test to screen for gestational diabetes. Gestational DM is managed with insulin injections. Although gestational DM usually resolves after delivery, a significant percentage of these women (perhaps 1 in 4) develop insulin-dependent diabetes years later.

Finally, DM can occur secondary to chronic treatment with certain medication, specifically glucocorticoids as well as calcineurin inhibitors (CNI). Long term use of oral corticosteroids e.g. prednisone produces Cushing's syndrome, marked by diabetes, hypertension, truncal obesity, abdominal striae, osteoporosis and poor wound healing. CNI drugs like Tacrolimus and Cyclosporine A increase one's risk of developing DM in addition to chronically damaging the organ allograft (esp. kidneys) they were meant to protect. Oral Cyclosporine has mostly fallen into disuse in light of its side effect profile: hypertension, elevated blood glucose, hyperlipidemia and increased risk of lymphoma and other cancers.

Essay 66: Diagnosis and Treatment of Emphysema

Emphysema is a form of chronic obstructive pulmonary disease (COPD), which also includes the disease state chronic bronchitis. Symptomatic emphysema occurs when 30% or more of the alveoli in a person's lungs have been destroyed. The lungs' total surface area for gas exchange is normally about the size of a tennis court. In contrast, people with severe emphysema have a surface area equivalent to a ping pong table available for gas exchange. The major symptoms of emphysema include wheezing, difficulty exhaling air from the lungs, shortness of breath (dyspnea) on exertion and ultimately at rest, as well as increased susceptibility to respiratory infections, particularly streptococcal pneumonia. The main risk factor for developing both forms of COPD is exposure to tobacco smoke. However, a rare form of emphysema is caused by a genetic disease known as alpha-1 antitrypsin deficiency (AATD).

Diagnosis

The onset of emphysema tends to be insidious; hence, many patients are not diagnosed until the disease has reached a fairly advanced stage. Most cases of emphysema are diagnosed in people over age 50, who generally have a long history of smoking. The main diagnostic techniques involve taking a careful history, performing a physical examination of the lungs, and obtaining chest X-rays and pulmonary function tests to confirm the suspected diagnosis.

In patients with emphysema, a chest X-ray shows an expanded thoracic volume as well as abnormally dark areas within the lungs. The dark regions signify the loss of lung parenchyma and subsequent air trapping within the cavities, or dead space, left behind. Pulmonary function tests focus on the volume of air a person can exhale and the time required to perform this action. Emphysema fits the classic pattern of obstructive lung disease: the FEV1 or forced expiratory volume of air in 1 second falls in relation to the FVC (forced vital capacity), i.e., the total volume of air expelled from the lungs with maximum effort.

In contrast, people with AATD tend to exhibit symptoms of emphysema by age 30. Often, these people are not smokers. In this particular variant of emphysema, a lack of the trypsin inhibitor AAT allows out-of-control protease enzymes to attack and ultimately destroy the alveoli in the lungs.

Emphysema symptoms in a young patient with no history of tobacco use should make any physician strongly suspect AATD. The diagnosis of AATD is confirmed by genetic testing. Treatment consists of injections of recombinant alpha-1 antitrypsin. Lung transplantation has been tried on occasion; unfortunately, most lung allografts are destroyed by the recipient's immune system within 3 to 5 years. In the near future, gene therapy may be an option for people with AATD.

Treatment

The main goal in patients with mild to moderate emphysema is to delay further disease progression. It is difficult to overemphasize the importance of smoking cessation in arresting, or at least slowing, the course of emphysema. A bronchodilator, usually an albuterol inhaler, is generally effective in relieving shortness of breath and wheezing. Inhaled corticosteroids are given for flare-ups of all stages of emphysema. Antibiotics are also given at the first sign of a respiratory infection.

Severe emphysema is treated with a long-acting beta-2 adrenergic agonist such as salmeterol or formoterol; a rescue inhaler (albuterol) for emergency relief; inhaled corticosteroids, and, in late stages of the disease, home oxygen.

Other adjunct treatments for emphysema include leukotriene receptor blockers like montelukast (Singulair) and the antimuscarinic drugs ipratropium (Atrovent) and tiotropium (Spiriva). As in AATD, lung transplants have been tried in patients with severe emphysema; in most cases, however, these patients experience graft rejection and/or opportunistic infections due to chronic immune suppression. In spite of the various treatments in use today, the only therapy known to prolong life in patients with late-stage emphysema is supplemental oxygen.

Essay 67: Different Strokes for Different Folks

Stroke, also known as cerebrovascular accident (CVA), is the fourth leading cause of death in the United States today. Approximately 800,000 Americans will suffer a stroke in any given year, with nearly 25% dying acutely or from subsequent complications over the following 12 months. This article will describe the pathogenesis of the major types of stroke and briefly discuss some treatment options.

Thromboembolic Strokes

85% of strokes result from a thrombus, or blood clot, blocking the flow of oxygenated blood to the brain. This scenario often occurs in the context of undiagnosed or undertreated atrial fibrillation. In atrial fibrillation, the abnormal fluttering movement of the heart's atrial chambers increases the chances of blood clotting on the internal surface of the atria or the mitral valve. If the resulting blood clot breaks off or embolizes, it can travel through the left ventricle into the aorta. From the aorta, the clot can enter a carotid artery, travel into the middle cerebral artery, and ultimately lodge in a branch of that artery.

Because the brain needs a continuous supply of oxygen, the neurons downstream of a blood clot malfunction and begin to die when their supply of oxygen is cut off. The tell-tale signs of stroke include headache, dizziness, confusion, blurred vision (especially in one eye), slurred speech, loss of balance or coordination, weakness (or paralysis) on one side of the body, loss of bladder or bowel control, and, in some cases, loss of consciousness.

Other cases of thromboembolic stroke occur when an atherosclerotic plaque in the wall of an internal carotid artery ruptures. The debris from the plaque serves as a nucleation point for clotting factors and platelets, quickly forming a blood clot. Arterial blood flow carries the blood clot into the middle cerebral artery, and an identical scenario ensues. Blood flow in the distribution of a branch of the left or right middle cerebral artery is obstructed, and the brain tissue supplied by that artery begins to die from oxygen deprivation.

If a person suffering from this type of stroke receives prompt medical attention, there is a good chance of avoiding permanent brain damage. First, a CT scan is performed to distinguish thromboembolic strokes from hemorrhagic strokes, discussed next. If a CT scan is negative for active bleeding, the patient is given an injection of a thrombolytic drug called tPA (tissue plasminogen activator), which dissolves blood clots. However, if blood flow is not restored within three hours of the onset of stroke symptoms, the brain damage is likely to be permanent.

Hemorrhagic Stroke

Approximately 13% of strokes occur in the wake of a blood vessel rupturing, an all too common complication of untreated high blood pressure. The blood vessel in question may be a branch of a middle cerebral artery, producing the symptoms described above. In other cases, the vessel may be located deep inside the brain, causing a so-called lacunar infarct, which can result in sensory or motor deficits (but seldom both). Hemorrhagic strokes can also result from head trauma or therapy with anticoagulant drugs such as warfarin. Currently, there is no specific treatment for most hemorrhagic strokes other than supportive care followed by occupational, physical, and speech therapy. In the appendix of this book, I propose a novel approach to acute stroke management.

The most catastrophic form of hemorrhagic stroke occurs when a weak, bulging region in a blood vessel wall called an aneurysm ruptures, leading to a subarachnoid hemorrhage (SAH).

Intracranial aneurysms are more common in people with polycystic kidney disease; still, the majority of cases are sporadic. People suffering from SAH complain of the worst headache of their lives and often experience seizures and loss of consciousness. Even with prompt hospitalization and surgical intervention, 30% of SAH victims do not survive.

Miscellaneous causes

The remaining cases of stroke arise from several causes, including encephalitis, brain tumors, untreated parasitic infections such as neurocysticercosis, and severe migraine headaches lasting longer than 72 hours. This last condition, called status migrainosus, is treated with serotonin receptor blockers like sumatriptan or, in certain cases, intravenous corticosteroids.

Essay 68: Five Commonly Misdiagnosed Diseases

One maxim taught in U.S. medical schools is "when you hear hoof beats, think horses, not zebras." This means when it comes to diagnosing illnesses, common conditions must be ruled out before rare conditions are considered. On one level, this makes sense; however, real life is seldom so simple. For all of the advances in modern medicine, diagnosis remains as much an art as a science.

Several factors contribute to errors in diagnosis: the symptoms of the disease are vague or similar to other diseases; the presentation of the illness is highly variable; or the disease occurs so infrequently that few physicians have seen cases of it outside of textbooks and journal articles. With that in mind, here are five medical conditions that often get misdiagnosed.

1) SLE (Systemic Lupus Erythematosus)

Although physicians are quite familiar with the classic presentation of lupus (a young woman with a butterfly-shaped facial rash, joint pain, and fatigue), many cases of this disease do not fit this profile. 10% of SLE cases occur in men, and case reports demonstrate that people of all ages and backgrounds can develop the disease.

The manifestations of SLE are highly variable, ranging from fever, skin rashes, and arthritis to life-threatening conditions like kidney failure, seizures, and stroke. Moreover, the onset of lupus is often insidious, making it difficult for a patient to remember exactly when s/he began to feel ill.

Finally, the diagnostic criteria of SLE (4 positive symptoms out of 11 possible categories) may delay a diagnosis for months to years. Not surprisingly, SLE has been misdiagnosed as rheumatoid arthritis, pleuritis, nephritis, multiple sclerosis, epilepsy, and rarely, as a psychiatric disorder.

2) Sarcoidosis

This relatively uncommon disease is marked by the formation of inflammatory structures called granulomas in almost any part of the body, but most often, the lungs.

For unknown reasons, sarcoidosis affects African-Americans and Scandinavians more often than other populations. Because pulmonary symptoms may be mild or non-specific, pulmonary sarcoidosis is easily confused with more common respiratory diseases like asthma or bronchitis. On chest X-rays, sarcoidosis may be misdiagnosed as tuberculosis or even metastatic cancer.

Extrapulmonary sarcoidosis can masquerade as virtually any other disease. Cases involving the nervous system, pituitary gland, endocrine system, bone marrow, spleen, skin, exocrine glands, heart, kidneys, liver, and gastrointestinal tract have all been described in the medical literature. It is also important to note that sarcoidosis is a diagnosis of exclusion, meaning all other conditions must be ruled out before a final diagnosis is made.

3) Multiple Sclerosis (MS)

Like SLE and sarcoidosis, this neurological disease is thought to be autoimmune in nature. For unknown reasons, B and T lymphocytes mount an assault on myelinated neuronal axons in the brain and spinal cord. On CT scan and MRI, the inflamed, demyelinated areas appear as bright spots called plaques. Although the classic presentation of MS is a woman of childbearing age with optic neuritis, over 1/3 of cases occur in men, and any part of the central nervous system can be affected.

As such, MS may be misdiagnosed as a stroke or brain tumor, especially if only a single area of inflammation is detected. MS may also be confused with the initial stages of other neurological diseases such as encephalitis or ALS (Lou Gehrig's Disease).

4) Pheochromocytoma

This tumor is believed to be the underlying cause of 1 in 1,000 cases of high blood pressure. Hence, it is an uncommon but not altogether rare condition. Pheochromocytomas usually arise from chromaffin cells in the adrenal medulla and produce excessive amounts of the hormones epinephrine (adrenaline) and norepinephrine. When the tumor cells release these catecholamines, a person may experience sweating, heart palpitations, feelings of anxiety, or even a fainting spell. In extreme cases, the presenting symptom of pheochromocytoma is a heart attack or stroke.

In most cases, however, the chief symptom of a pheochromocytoma is hypertension that does not respond to multiple medications. Considering the wide variety of antihypertensive drugs on the market and patient non-compliance due to side effects or simple forgetfulness, this tumor may be misdiagnosed as essential hypertension, an anxiety disorder, or cardiac disease.

5) Atypical heart attack or silent MI (myocardial infarction)

Of all the conditions discussed in this article, an atypical or silent MI is by far the most common. Even experienced physicians can misdiagnose a silent MI (or fail to detect one entirely). One reason is that patients with an atypical heart attack may complain of vague aches and pains as opposed to the classic presentation of crushing substernal chest pain radiating to the left arm, sweating, and shortness of breath. Sometimes, the only symptom may be jaw pain, which can easily be misdiagnosed as a toothache, or shoulder pain, which is often dismissed as arthritis. Some patients complain of malaise, dizziness, or a feeling of lightheadedness. These symptoms can easily be mistaken for a viral infection, simple dehydration, hypoglycemia, and a host of other ailments. Postmenopausal women, the elderly, and people with long-standing diabetes are more likely than others to suffer an atypical heart attack.

Essay 69: Foods that Help Prevent Heart Disease

In order to understand the basis of heart-healthy foods, it is first necessary to understand the environmental and dietary factors that promote atherogenesis ("hardening of the arteries") and, by extension, heart disease. First, compounds that induce free radical formation, such as tobacco smoke and charbroiled meat, are considered culprits. Both contain polycyclic aromatic hydrocarbons, which not only increase one's risk of cancer but also cause vascular damage, a key event in the pathogenesis of heart disease.

Second, a diet high in saturated animal fats is also considered atherogenic. Saturated fats are believed to promote the formation of inflammatory autacoids (especially prostaglandin E2, thromboxane A2, and certain leukotrienes) that damage vascular endothelial cells directly and activate certain white blood cells, further exacerbating the damage. Foods rich in cholesterol, such as beef liver, may also contribute to this process if eaten in excess. Finally, a diet low in fiber increases intestinal transit time, allowing more fatty acids and cholesterol to be absorbed from the intestine and ultimately into circulation.

Heart-healthy foods minimize or even combat the above-mentioned effects. Foods rich in antioxidants like vitamins C, E, and lycopene include cranberries, pomegranates, citrus fruits, sunflower seeds, and cooked tomato products. Green tea also contains antioxidants. Some researchers speculate that its cardioprotective effects account for the lower rate of heart attacks in Japanese men who smoke cigarettes compared to their American counterparts.

The second group of heart-healthy foods is the so-called cardioprotective fats - unsaturated oils such as olive and canola oils and fish rich in omega-3 fatty acids. These fish include salmon, tuna, sardines, and cold water fish such as cod, herring, and haddock. The theory goes that omega-3 fatty acids promote the formation of less inflammatory prostaglandins, thromboxanes (PGE3 and TXA3), and prostacyclins, which have anti-inflammatory effects.

A professor of mine once noted that the rate of heart disease in the United States increased during the same time frame that fish consumption declined.

Although this explanation seems somewhat simplistic (for example, it ignores the dramatic increase in tobacco use during the 20th century), the dearth of omega-3 fatty acids in the American diet may well contribute to the high prevalence of heart disease in the U.S.

Finally, a high-fiber diet (whole grains, legumes, prunes) has many positive effects, such as bowel regularity and, probably, a decreased risk of colon cancer. With regard to cardiac health, dietary fiber binds bile acids, which are cholesterol derivatives. This promotes their disposal from the body as opposed to their reuptake by the liver, which then diverts cholesterol production to replace bile acids rather than produce LDL cholesterol (the so-called "bad cholesterol"). The overall effect is less cholesterol circulating in the bloodstream, which decreases vascular endothelial cells' exposure to this atherogenic substance.

Although research will undoubtedly continue, it seems that antioxidants, omega-3 fatty acids, and plenty of fiber are the cornerstones of a heart-healthy diet.

Essay 70: Hashimoto's Thyroiditis

Hashimoto's thyroiditis is an autoimmune disease marked by the partial or total destruction of the thyroid gland by B and T lymphocytes. As with other autoimmune disorders, the cause/s of Hashimoto's disease remains unknown; however, the epidemiology is well described. The prevalence of Hashimoto's may be as high as 1 in 20 people in the U.S. over the course of a lifetime. Women are diagnosed with this condition five to ten times as often as men. Most cases of Hashimoto's occur in middle age, although occasional cases have been reported in adolescents and children.

Signs and Symptoms

The onset of Hashimoto's thyroiditis tends to be insidious. Over the course of months, patients often notice a goiter along with the telltale signs of an underactive thyroid gland: cold intolerance, involuntary weight gain, coarse hair, and fatigue. Sometimes, inflammation of the thyroid causes a deepening of the voice or chronic hoarseness. A slow heart rate (bradycardia) is also common. Many patients report feelings of depression as well.

Laboratory findings

A series of laboratory tests are used to confirm the diagnosis of Hashimoto's. These include a TSH level (thyroid stimulating hormone) and levels of T3 and T4, the two main hormones produced by the thyroid gland. T3 and T4 levels are low in hypothyroidism, whereas TSH is elevated. Essentially, the pituitary gland senses the lack of thyroid hormone and kicks into overdrive, churning out high levels of TSH in an ultimately futile attempt to coax the failing thyroid gland to produce more T3 and T4. Another laboratory finding specific to autoimmune thyroiditis is the presence of antithyroglobulin antibodies, also called anti-TPO or anti-thyroid peroxidase antibodies. Finally, people with hypothyroidism often exhibit abnormal lipid profiles, especially elevated levels of cholesterol (possibly due to slower liver metabolism).

Treatment

In most cases of Hashimoto's, the thyroid gland succumbs to the autoimmune assault. Therefore, therapy aims to replace the hormones normally produced by the thyroid gland. Thyroid hormone is marketed generically as levothyronine or levothyroxine, better known by its trade name Synthroid.

Replacement of thyroid hormones almost always reverses the manifestations of hypothyroidism. It is important to note that untreated hypothyroidism usually becomes progressively worse over time. In extreme cases, the end result is a fatal condition called myxedematous coma. Even if this worst-case scenario does not ensue, long-term complications include heart and respiratory failure. As such, patients with permanent hypothyroidism (as a result of Hashimoto's, radiation exposure, and surgery) are advised to follow up with a physician at least every 12 months to ensure their medication dosage is effective.

Essay 71: Heat Stroke

A heat stroke is a medical emergency when a person's body overheats to the point that its normal mechanisms for dissipating heat fail. Core body temperature rises dangerously high, sometimes to over 42 degrees Celsius or 106 degrees Fahrenheit. Unless a person suffering from heat stroke is treated promptly, the stage is set for potentially fatal events, namely cardiac arrhythmias, seizures, and even coma due to widespread brain damage.

As with all worst-case scenarios, heat strokes tend to occur gradually rather than suddenly. When the human body is exposed to hot temperatures, sweating begins in a minute or less. Sweating is a generally effective strategy for removing heat from the body by convection. Sweat glands secrete a dilute, somewhat salty liquid that quickly evaporates, cooling the skin beneath it in the process. Compared to radiation or conduction convection is the most effective way of moving heat compared to radiation or conduction.

The trouble begins when a person becomes dehydrated faster than s/he replaces the fluid lost as sweat. Serum electrolytes, especially sodium, and potassium, also become depleted with excessive sweating. At this stage, termed heat exhaustion, people complain of feeling weak, dizzy, disoriented, and sometimes nauseated. People suffering from heat exhaustion show signs of fatigue and profuse sweating but usually quickly recover once they drink orange juice or, alternatively, ingest water with salt tablets. An essential point to remember is that not only water but also salt must be replaced; otherwise, dangerous electrolyte imbalances can result.

Ignoring the signs of heat exhaustion is a recipe for disaster in the form of heat stroke. The most ominous sign of heat stroke is the absence of sweating. The person's lips become dry and cracked, and their skin appears red and warm. Ironically, the victim may complain of feeling cold rather than hot. By this stage, core body temperature is entering the danger zone, approximately 41 degrees Celsius or 105.8 degrees Fahrenheit. If the person is not cooled quickly, seizures and/or cardiac arrhythmias are likely to ensue.

Populations at higher risk of suffering a heat stroke include children, the elderly, people who use saunas for an extended length of time, outdoor laborers, athletes who wear full uniforms in hot weather, and people with certain medical conditions, including thyrotoxicosis and Down's Syndrome.

A person suffering from heat stroke should be moved out of direct sunlight, preferably to an indoor setting with air conditioning, and stripped of clothing. The recommended method for cooling the victim is to apply ice packs to as much of the skin surface as possible until sweating resumes. If possible, the ice or frozen material should be wrapped in plastic to avoid skin damage. A cold shower or bath may be used if ice or frozen packs are unavailable. After this step is underway, rehydration should be initiated orally if the victim is able to swallow or intravenously if the victim is unresponsive and the appropriate supplies are available. As a reasonable precaution, heat stroke victims should be transported to a hospital or other medical facility as soon as possible.

Essay 72: Huntington's Disease

Huntington's disease (HD) is a central nervous system disorder marked by a progressive loss of motor control and involuntary upper body movements (sometimes called chorea because these tics vaguely resemble dancing), followed by autonomic dysfunction, dementia, and eventually death. In most cases, death occurs 10-15 years after the onset of symptoms. The age of disease onset varies but tends to occur after a person has reached reproductive age. HD is an autosomal dominant disorder. This means that anyone with a single copy of the abnormal gene responsible for HD has a 50% chance of passing the disease on to his or her offspring.

Although the hereditary pattern of HD has been understood for nearly a century, the genetic mutation underlying the disease was not identified until 1993. That year, researchers studying a large Venezuelan family with numerous cases of HD finally sequenced the stretch of DNA on chromosome 4 containing the abnormal HD gene. The gene in question encodes a protein named Huntingtin in honor of the physician after whom the disease was originally named.

Soon after the huntingtin gene was identified, geneticists compared its sequence in HD patients to that of healthy controls. They discovered that healthy people have 35 or fewer repeats of a trinucleotide sequence called CAG in their huntingtin genes. CAG encodes the amino acid glutamine. In HD patients, however, the huntingtin gene contains an expansion of the CAG repeat to 36 or more copies, resulting in an unusually long stretch of glutamines in the resulting protein. These glutamines, in turn, seem to confer an abnormal gain of function on the huntingtin protein, ultimately disrupting cellular metabolism to the point of causing cell death.

Unlike autosomal recessive diseases like sickle cell anemia or Tay-Sachs, which result from two defective alleles, a single copy of the mutated huntingtin gene is sufficient to cause full-blown Huntington's disease. Apparently, one normal copy of the huntingtin gene cannot compensate for the presence of the mutant gene. Furthermore, the higher the number of CAG repeats, the earlier the onset of HD symptoms; in rare cases, HD can strike in adolescence.

Several features of HD remain a mystery. First, no one has figured out the normal physiological role of the huntingtin protein in humans. Only subtle behavioral changes result when the corresponding gene is deleted in knockout mice, and the mouse's lifespan is unaffected.

Second, it is unclear why HD strikes a specific region of the brain called the basal ganglia; after all, every cell in a person's body, including the nervous system, contains the abnormal huntingtin gene. Finally, no one has offered a convincing explanation as to why the onset of HD and other neurological diseases caused by CAG repeat expansions is usually delayed until adulthood.

Treatment of HD is largely supportive. Antidepressants, mood stabilizers, and dopamine receptor blockers often alleviate psychiatric symptoms. Coenzyme Q10 may slow the course of the disease, but its efficacy is still a matter of debate, as is its mechanism of action.

People with a family history of HD face the awful dilemma of undergoing genetic testing. Some prefer definitive diagnosis years in advance. Proponents of genetic testing argue that if a test is done early enough, people with HD can use this knowledge to decide whether or not to have children.

Opponents cite the emotional impact of potentially devastating news. They argue that people may regard an early diagnosis of HD as a virtual death sentence, leading to severe depression and even suicide. As with many other diseases for which genetic tests exist but effective treatments do not, there are no easy answers.

Essay 73: Pathogenesis of Primary Hypertension

Hypertension (HTN), commonly known as high blood pressure, induces pathological changes throughout the cardiovascular system, particularly in the body's arteries. One reason arteries suffer a disproportionate amount of damage stems from the human circulatory system's basic anatomy: The heart's left ventricle generates far more contractile force than the heart's other chambers.

The arterial system constitutes the high-pressure side of circulation; consequently, when the left ventricle ejects blood into the aorta, its walls are exposed to a surge of pressure with each heartbeat. In chronically elevated blood pressure, the large arteries can accommodate the stronger pressure waves due to their thick walls; smaller arteries, however, undergo the pathological process known as atherosclerosis. More on that in a moment.

Years of research have revealed that the heart is only part of the story. It seems that the sympathetic nervous system, kidneys, smooth muscle, and endothelial cells lining the blood vessels all contribute to the dysregulation of vascular tone, culminating in hypertension. Surprisingly, the cause(s) of 90% of HTN cases remains obscure. Risk factors include family history, obesity, smoking, and, in some cases, excessive sodium intake. Regardless of HTN's etiology, its chronic effects are well understood.

Medium-sized arteries seem to be the blood vessels most prone to developing atherosclerosis. In the popular press, this term is defined as "hardening of the arteries," a largely accurate description. Basically, when arterial walls are subjected to chronically elevated pressure, cells called fibroblasts proliferate within the intimal layer of the arterial wall; a process called intimal thickening. Comorbid conditions like high cholesterol and diabetes further accelerate atherosclerosis.

Over time, the lumen of these vessels becomes progressively narrower, impairing blood flow in many parts of the body, such as the lower extremities, the coronary arteries supplying the heart muscle, and the carotid arteries, which supply blood to the brain. Not surprisingly, this sets the stage for heart attacks and strokes, two of the leading causes of death in the Western world today. Of course, poor circulation in the legs and feet is a setup for gangrene and amputation.

In contrast to medium-sized arteries, small arteries, and capillaries are prone to hypertensive damage in the form of vascular retinopathy and lacunar infarcts. Simply put, chronic HTN wreaks havoc on the retinal blood vessels. Ophthalmologists have coined descriptive terms for these pathological changes, including cotton wool spots, copper wires, and proliferative retinopathy, most often seen in diabetics.

In addition to retinal damage and blindness, a second consequence of long-standing HTN on small vessels is a phenomenon called lacunar infarcts. These can be thought of as microvascular ruptures or mini-strokes, which often occur in a region of the brain called the basal ganglia, producing distinct motor deficits, including expressive aphasia or difficulty with motor control of speech.

Chronic Kidney Disease (CKD)

No discussion of HTN-induced vascular damage would be complete without mentioning chronic kidney disease. Long-standing HTN causes irreversible damage to the renal glomeruli. Over the course of several years, the protein network surrounding the glomerular blood vessels called the basement membrane, becomes scarred with collagen and other fibrotic debris.

The blood filtration rate through the kidneys, the glomerular filtration rate or GFR, declines as the nephrons gradually lose their ability to filter plasma into urine. Kidney failure ensues when 75% or so of the renal nephrons become non-functional.

Unfortunately, the kidneys respond to this decline in GFR by activating a protein called angiotensin. This protein raises blood pressure in three ways: 1) direct vasoconstriction; 2) production of the hormone aldosterone, which promotes sodium retention; and 3) activating hypothalamic neurons responsible for the sensation of thirst. This creates a vicious cycle in which the failing kidneys hasten their own demise by continually raising arterial blood pressure. Ultimately, patients with chronic renal failure require dialysis or a kidney transplant to remain alive.

Essay 74: Hypertension – Secondary Causes

As noted in the previous essay, approximately 90% of hypertension cases are termed primary or essential hypertension, meaning that no comorbid disease or identifiable process is responsible for the elevation in blood pressure. In 10% of cases, however, elevated blood pressure occurs secondary to some other disturbance, most often a vascular malformation, endocrine disorders, or certain types of tumors. Secondary hypertension is often termed "resistant" because the person's blood pressure remains elevated above the healthy range of 120/80 mmHg despite numerous medical interventions. This article will briefly explore some causes and characteristics of secondary hypertension.

Adrenal tumors

One form of adrenal cortical tumor, called Conn tumor, releases large amounts of the hormone aldosterone. This hormone, the body's major mineralocorticoid, promotes sodium retention in the kidney and raises blood pressure. Another tumor of the adrenal cortex causes a constellation of signs and symptoms called Cushing syndrome. A related condition called Cushing's disease results from an ACTH-releasing tumor in the anterior pituitary gland. Patients with Cushing's suffer from obesity, diabetes, osteoporosis, and hypertension.

The adrenal medullary tumor known as pheochromocytoma can also produce resistant hypertension. These tumors arise from chromaffin cells in either the adrenal medulla or, occasionally, in the paravertebral ganglia. 10% of pheochromocytoma cases occur bilaterally; 10% are malignant; and 10% occur in the context of genetic disorders, including Von Hippel-Lindau, Neurofibromatosis Type I, and Multiple Endocrine Neoplasia Type II.

Most adrenal tumors can be removed surgically. In rare cases of inoperable pheochromocytoma, adrenergic receptor blockers like phentolamine and phenoxybenzamine are given. Cases of hyperaldosteronism not caused by a detectable tumor can be treated with the aldosterone receptor antagonist spironolactone.

Vascular malformations

The most common one underlying resistant hypertension is coarctation of the aorta. People with this condition may be asymptomatic (aside from hypertension). They often exhibit a wide discrepancy in blood pressure when readings are obtained from both arms. Aortic coarctation is a surgically correctable condition.

Thyroid disorders

Hyperthyroidism, e.g., Grave's disease, can also lead to resistant hypertension. This generally occurs if the thyroid hormone imbalance goes undiagnosed or the person delays seeking treatment. The hallmark of this condition is a wide pulse pressure, meaning an elevated systolic blood pressure but a normal diastolic pressure. Once the hyperthyroidism is corrected, the person's blood pressure usually returns to normal.

Drug-induced hypertension

This category is often overlooked even by physicians. A number of medications can induce hypertension as a side effect. The most notorious culprits are chronic treatment with corticosteroids, producing a syndrome akin to Cushing's; the hormone erythropoietin, used to combat anemia; and the anti-rejection drugs Cyclosporine A and Tacrolimus (FK506), used in organ transplant recipients.

Essay 75: Hypertension: Non-Pharmacological Treatment Options

Hypertension (HTN), better known as high blood pressure, is one of the most common health problems in the United States today. According to the American Heart Association, an estimated 74 million Americans are believed to have hypertension, while millions more have "borderline" hypertension, a condition that generally progresses to hypertension over time. According to the most recent JNC (Joint National Committee) definition, a person is considered to have some form of hypertension if, on three separate occasions, his or her resting blood pressure is greater than or equal to 140 mmHg (millimeters of mercury) systolic and/or 90 mmHg diastolic.

Approximately 90% of people with HTN have the so-called essential form of hypertension, meaning no underlying condition is linked to the sustained rise in blood pressure. The remaining 10% of HTN cases are secondary, arising from any number of causes, including anatomic abnormalities like aortic coarctation; endocrine disorders like hyperthyroidism and adrenal tumors; a side effect of various medications, especially steroids and oral contraceptives; and, of course, adverse drug interactions.

Countless volumes have been written about the medical treatment of hypertension; however, the remainder of this article will focus on ways to lower blood pressure without resorting to antihypertensive drugs. As a disclaimer, this approach is NOT guaranteed to reduce anyone's blood pressure to normal levels, and people with hypertension are strongly urged to follow up with a physician on a regular basis. That said, all concerned will agree that the measures discussed below are highly beneficial to a healthy lifestyle.

1) Smoking cessation. If you smoke or use any form of tobacco, do whatever it takes to quit. If that requires you to join an ex-smoker support group or undergo hypnosis therapy, so be it. (Nicotine patches and gum technically count as drugs, but one can argue that they are indicated for smoking cessation, not to treat high blood pressure.) In any event, if you smoke, quit as soon as possible.

2) A healthy diet. You've heard the drill umpteen times. Eat plenty of fruits, vegetables, and whole grains.

Avoid refined sugar and saturated fats. Lean cuts of red meat are occasionally fine, but poultry and fish are preferable. In some cases of hypertension, a low-sodium diet is also beneficial. It is unclear how many patients have salt-sensitive hypertension and processed foods' high sodium content only worsens matters. So try to avoid processed foods as much as possible. As the saying goes, eat to live; don't live to eat. And when in doubt, put down the salt shaker.

3) Red wine is good in moderation (1 glass daily); excess alcohol in any form has negative health consequences. Once again, moderation is crucial.

4) Exercise and an active lifestyle. Everyone knew this one was coming. Maintaining a healthy weight is good for you. There is literally no downside to staying in shape, and not much point in reading a lengthy sermon about the myriad dangers of obesity.

5) Relaxation techniques. Stress reduction techniques include meditation, biofeedback, and yoga. All three may be the functional equivalents of self-hypnosis; in any case, they seem beneficial, so trying them is not harmful.

6) A sleep evaluation. Anyone who suspects they may have obstructive sleep apnea (OSA) will benefit from an overnight evaluation in a sleep laboratory. Excessive snoring and daytime sleepiness are red flags for OSA. Treatments include adhesive nasal strips for mild cases, CPAP (Continuous Positive Airway Pressure) machines for moderate to severe cases, and the option of surgery for extremely severe cases. In any case, addressing sleep apnea often helps reduce hypertension.

Essay 76: Leprosy

Hansen's disease is a largely obsolete term for the infectious condition better known as leprosy. Leprosy is a bacterial infection thought to affect some 3 million people worldwide, with over 300,000 new cases identified annually. The signs and symptoms of leprosy are highly variable. Some patients develop pale, discolored patches of skin. In other cases, the hallmarks are pain, tingling, or numbness of the extremities. In the most severe cases, patients suffer facial disfigurement, blindness, and extensive peripheral nerve damage, leading to the loss of fingers or toes.

The infectious agent in leprosy is a mycobacterium called M. leprae, whose cell structure and reproductive cycle resemble those of the bacterium responsible for causing tuberculosis (M. tuberculae). Unlike TB, however, very few cases of leprosy have been diagnosed in the United States in the past 50 years. Today, the vast majority of leprosy cases occur among the rural poor in tropical countries, especially Bangladesh, Brazil, India, Mozambique, Myanmar, and Somalia.

Diagnosis

Although leprosy can infect people of any age, children seem to be more susceptible than adults. Patients infected with M. leprae exhibit one of two major disease manifestations: tuberculoid leprosy or lepromatous leprosy. Considering that mycobacteria are not known to produce any toxins, the variable symptoms of leprosy reflect differences in how an individual's immune system responds to the bacteria. Briefly, tuberculoid leprosy shows up as patches of skin numbness and discoloration but tends to run a mild course. On the other hand, patients with untreated lepromatous leprosy often suffer severe complications, including facial disfigurement, muscle weakness, and progressive nerve damage leading to chronic limb injuries and the eventual amputation of fingers and toes.

M. leprae cannot be cultured on an agar plate, unlike many other bacteria. Instead, it must be grown on the footpads of armadillos. Since this technique is only feasible in research labs, the traditional method of diagnosing leprosy is to stain skin lesion scrapings for Acid Fast Bacilli (AFB). The major drawback is that AFB staining may not reveal the presence of any mycobacteria in a patient's skin lesions, especially in cases of advanced disease.

An alternative diagnostic method was devised in the 1990s using PCR (Polymerase Chain Reaction) to amplify mycobacterial DNA or ribosomal RNA sequences. Theoretically, PCR is sensitive enough to detect the presence of a single mycobacterium. Unfortunately, PCR thermocyclers are few and far between in the regions of the world where most cases of leprosy occur. As such, if a physician suspects leprosy based on clinical presentation.

Treatment

The mainstay of leprosy treatment in much of the world is a triple-drug regimen consisting of dapsone, clofazimine, and rifampicin. Dapsone works by blocking bacterial folate synthesis and, by extension, bacterial reproduction. Clofazimine is believed to damage bacterial DNA by generating toxic oxygen radicals. Rifampicin works by inhibiting an enzyme called RNA polymerase, effectively shutting down bacterial gene transcription. According to World Health Organization guidelines, these medications must be taken anywhere from 6 to 24 months to ensure adequate treatment of leprosy. Other anti-inflammatory medications, such as aspirin and prednisone, are often used in conjunction with this regimen.

Several factors conspire to make effective treatment of leprosy difficult. First, there is the obvious problem of identifying new cases of the disease in a timely manner.

The asymptomatic phase of leprosy may last 3 years or longer, making it virtually impossible to treat patients or their close contacts prophylactically. Second, although the antibiotics used to treat leprosy are not particularly expensive, relatively few pharmaceutical companies manufacture them in light of their rare use in the U.S. and Europe. Third, as noted above, a course of drug therapy can last up to 2 years, making follow-up care extremely difficult in remote rural areas. On that note, if patients fail to complete a full course of therapy, antibiotic-resistant strains of leprosy can emerge, making the infection impossible to eradicate.

Why Leprosy Is (Probably) Here to Stay

In the realm of infectious disease, the success or failure of eradication efforts comes down to etiology (mechanisms of disease causation), available treatment options (vaccines and drugs), and epidemiology (affected populations). Leprosy presents obstacles in all three categories. Although leprosy is far less contagious than TB, surprisingly, little is known about the disease's spread or pathogenesis. Captive armadillos are capable of carrying leprosy, although virtually no human cases have been linked to contact with wild animals. This leaves infected humans as the main reservoir of leprosy bacteria; nevertheless, the exact mode of disease transmission remains obscure.

Along with its slow spread, the insidious course of leprosy makes timely detection of new cases nearly impossible. External lesions and skin discoloration take months or years to develop. In other words, most people with leprosy show no outward signs or symptoms long after the initial infection.

In addition, people infected with leprosy can survive for decades; the disease is often debilitating but rarely fatal. Diseases that fit this profile a long incubation period followed by an indolent course – are seldom, if ever, completely eradicated.

The chronic nature of leprosy necessitates a protracted treatment course. Existing drug regimens can cure most cases of leprosy; however, these regimens consist of three or more drugs, and the course of treatment may last over 1 year. A lengthy treatment course is a recipe for a high rate of treatment failure. In the case of TB, it also accounts for a sharp rise in antibiotic-resistant strains of the bacteria. Although comparatively few reports of antibiotic-resistant strains of leprosy have surfaced in the medical literature, this phenomenon has occurred innumerable times with other bacteria. It is naïve to expect leprosy to be an exception to the rule.

Next, there are key epidemiological factors to consider. Leprosy is largely a disease of the poverty-stricken in the developing world. Most cases of leprosy occur in tropical countries, especially Brazil, India, Bangladesh, the Congo Basin, Mozambique, Somalia, and the Indonesian archipelago. This factor alone explains why eradication efforts remain so difficult. Leprosy is not on the radar screen of most wealthy countries, namely the United States, Western Europe, Japan, or even China. As the saying goes, out of sight, out of mind.

As a corollary, only a few pharmaceutical companies manufacture the drugs necessary to treat leprosy, such as rifampicin, dapsone, or clofazimine. The reasons are obvious: Except for rifampicin, these drugs are rarely used in the industrialized world. For example, clofazimine is unavailable in the U.S. except under a single HSRA research protocol through the FDA. Second, from the manufacturers' perspective, there is little, if any, profit to be made from the sale of these drugs. Unless governments, international organizations, or private charities subsidize the production of these medicines, pharmaceutical companies will have no incentive to produce them.

Finally, relatively few physicians in wealthy countries specialize in tropical diseases. Although this trend is slowly changing, physicians tend to concentrate their efforts on highly transmissible diseases with a short treatment course and a high cure rate, like malaria. Meanwhile, the countries with the highest number of leprosy cases tend to have a shortage of physicians in general and relative to the rural population in particular. Most physicians choose to live in urban or suburban areas for social, economic, and practical reasons. This trend is unlikely to change in the near future, further confounding attempts to eradicate leprosy at the global level.

Essay 77: Treatment Options for Drug Resistant Leukemia

Drug-Resistant Leukemia?

As with many other forms of cancer, a major obstacle to the successful treatment of leukemia is the emergence of cancer cells resistant to first-line chemotherapy agents. In other words, the drugs used in the initial round of chemotherapy may fail to induce remission altogether, or drug-resistant cancer cells reappear after an interval of months to years. In both scenarios, treatment options are limited but fall into three general categories: a) alternative chemotherapy agents to which the patient has not been exposed, b) a trial of immunotherapy, or c) a bone marrow or stem cell transplant as a treatment of last resort.

Alternative chemotherapy agents

Since the approval of Imatinib (Gleevec) for the treatment of CML (chronic myelogenous leukemia) over a decade ago, tyrosine kinase inhibitors have come into widespread use for the treatment of other forms of leukemia as well as several other cancers during a recurrence. The rationale behind tyrosine kinase inhibitors is that they block growth-promoting signals from a variety of sources: overexpressed growth factor receptors on the cell membrane, Src family tyrosine kinases in the cytosol, as well as the abnormal BCR-ABL protein that drives proliferation of leukemia cells in CML and 30% of cases of ALL (acute lymphoblastic leukemia). Because tyrosine kinase inhibitors bind to multiple target proteins, chances are the tumor cells will maintain some degree of susceptibility to them even as they become resistant to other chemotherapy agents.

Immunotherapy

Briefly, immunotherapy involves extracting a patient's lymphocytes (assuming healthy ones can be found), exposing them to tumor antigens, and then injecting these primed lymphocytes back into the patient. The hope is that following this special round of training, the lymphocytes will destroy as many tumor cells as possible. This strategy has shown promise with melanoma; whether it works effectively in leukemia patients remains to be seen.

A second strategy that shows promise is the development of monoclonal antibodies to target tumor cells selectively. One example is Rituximab (Rituxan), first used in the treatment of CD20 positive Non-Hodgkin's Lymphoma (NHL). Although NHL differs in some respects from leukemia, both forms of cancer tend to arise as tumors of B cell origin. As of 2010, Rituxan was approved for use in treating CLL (Chronic Lymphoblastic Leukemia).

Bone marrow and Stem Cell Transplants

Bone marrow transplants (BMT) are sometimes successful in treating recurrent leukemia. The strategy is to extract bone marrow from a genetically compatible donor, subject the patient to total body irradiation, and then use this marrow (which contains the stem cell progenitors of every kind of blood cell) to reconstitute the patient's immune system with healthy cells. The acute risks are that the transplanted cells will not take up residence in the patient's marrow or else fail to produce enough functional blood cells to reconstitute the patient's immune system. In either case, the invariable outcome is death from opportunistic infections. A long-term complication of BMT is a phenomenon called Graft Versus Host Disease (GVHD), described below.

In recent years, stem cell transplants have superseded BMT at various tertiary care centers across the U.S. The overall procedure remains the same; however, the advantage of using purified hematopoietic stem cells over bone marrow is a much lower risk of GVHD.

This disorder results from the introduction of donor lymphocytes (or their precursors), which regard the recipient's entire body as foreign tissue (as opposed to self). Weeks or even months after the transplant, these autoreactive lymphocytes attack the skin, liver, and/or gastrointestinal tract. GVHD is potentially fatal; hence, oncologists tend to prefer stem cell transplants over BMT in the treatment of leukemia.

Many cases of Non-Hodgkin's Lymphoma share features with ALL and CLL leukemia, in particular their B cell origin. A new drug was approved in 2020 called Tazemetostat (Tazverik) for a subset of relapsed follicular NHL cases. This drug is given orally twice a day and works by blocking the methylation of H3K27 - the lysine at position 27 of histone subunit 3. Inhibiting methylation is thought to have a two pronged effect. First, the promoters of certain pro-apoptotic genes are exposed to RNA polymerase, effectively activating the programmed cell death pathway. Second, when enough histones remain demethylated, chromatin cannot condense properly during mitosis. Instead of condensed chromosomes lining up in an orderly fashion, the cancer cell's chromosomes become a hopelessly tangled mess. As the tangled chromosomes pull apart during anaphase, mitotic catastrophe ensues and the daughter cells do not survive.

Essay 78: Lung Cancer – Diagnosis, Staging, and Treatment

In this essay, I have chosen to focus on the clinical aspects of lung cancer diagnosis as opposed to an overemphasis on epidemiology. As of 2011, lung cancer is the leading cause of cancer deaths in both men and women in the U.S., which translates into over 150,000 deaths per year. Over 90% of lung cancer cases occur in people with a history of tobacco use, secondhand smoke exposure, or occupational exposure to asbestos. Despite recent advances in the diagnosis and treatment of cancer, the 5-year survival rate for lung cancer continues to hover at a disappointing 15%. The remainder of this article will steer clear of statistics.

Diagnosis

The following symptoms should be considered red flags by all healthcare providers.

1) Hemoptysis. In layman's terms, it means coughing up blood. Once other obvious causes like tuberculosis have been ruled out, the possibility of lung cancer must be considered.

2) Constitutional symptoms include fevers, night sweats, weight loss, and malaise in any middle-aged or older person with a history of tobacco use.

3) Other non-specific pulmonary symptoms: a persistent cough, worsening shortness of breath, or frequent respiratory infections, especially pneumonia.

4) Less common signs and symptoms that warrant a workup for lung cancer include pathologic bone fractures in a person not known to have osteoporosis and the so-called neoplastic and paraneoplastic syndromes.

5) Horner's syndrome - this constellation of findings consists of eyelid droop (ptosis), pupillary constriction (miosis), and decreased sweating (anhydrosis) on one side of a person's face. Lung cancers that compress a bundle of nerves called the brachial plexus are an important cause of Horner's syndrome. Because Horner's Syndrome directly results from the invasive spread of cancer into the brachial plexus, it is termed a neoplastic syndrome.

6) A well-documented paraneoplastic syndrome is a condition called Lambert-Eaton Myasthenia Syndrome, or LEMS for short. Patients with LEMS develop worsening muscle weakness and motor deficits due to auto-antibodies (allegedly in response to tumor proteins) that bind to and inactivate voltage-gated calcium channels on neurons.

The diagnosis of lung cancer generally involves a combination of imaging (chest X-ray, CT, MRI), bronchoscopy, and biopsy. If a distinct mass is detected on an X-ray film or by CT or MRI, the next step is to visualize the mass using a bronchoscope. This step makes sense if the tumor is located near the large airways or main stem bronchi. If the tumor involves the lung parenchyma, meaning the small airways and alveoli, the next step is usually a biopsy.

A detailed description of a lung biopsy procedure is beyond the scope of this article. Briefly, samples of the tumor may be obtained by CT-guided needle aspiration, thoracentesis of a fluid pocket, and/or collection and analysis of BAL (bronchoalveolar lavage) fluid.

Staging and Treatment

Squamous cell lung cancer arises from the epithelial cell layer lining the airways. Histologically, its appearance is distinct from adenocarcinoma, which tends to have a more glandular appearance; bronchoalveolar carcinoma, often similar in appearance to normal lung tissue; and small cell lung cancer, which arises from neuroendocrine cells and is sometimes called oat cell carcinoma.

For the purposes of staging and treatment, lung cancers are divided into two categories: small-cell and non-small-cell lung cancer. Small cell lung cancer (SCLC) is almost never treated surgically, owing to its tendency to have spread extensively by the time a person is diagnosed. Many cases of SCLC respond to radiation and chemotherapy drugs. Nonetheless, the median survival rate of SCLC remains at about one year. In contrast, non-small cell lung cancer (NSCLC) may be surgically resected when the disease is localized to one lung or the lobe of a lung. Radiation and chemotherapy (often involving platinum-based agents) are used as well.

Unfortunately, over 2/3 of patients with NSCLC have extensive disease at the time of diagnosis, resulting in a median survival rate of 2 years or less.

Essay 79: Neuroblastoma

Neuroblastoma is a solid tumor of the sympathetic nervous system that occurs almost exclusively in children under 2 years old. As many as 1/3 of neuroblastomas arise in the adrenal medulla, whereas the remainder may arise anywhere along the sympathetic chain, a series of nerve ganglia that extends from the thoracic spine to the lower abdomen. Histologically, the tumor most often appears as sheets of undifferentiated, small, round, blue cells. In a small number of cases, the neuroblastoma cells exhibit some degree of differentiation and resemble peripheral nervous tissue.

The most common manifestations of neuroblastoma include a palpable abdominal mass, fever, and weight loss, especially in children under 12 months old. Older children may not exhibit signs of the disease until the tumor has spread to the eyes, skin, liver, lungs, or bones. Approximately 500 new cases of neuroblastoma are diagnosed annually in the United States. According to an NIH estimate, approximately 1 child in 100,000 will be diagnosed with this tumor.

Diagnosis and Treatment

In cases of suspected neuroblastoma, imaging studies are done first, including X-rays, MRI, and sometimes a nuclear medicine study called an MIBG scan to confirm the presence and location of the tumor. Once the primary tumor has been located, a surgical biopsy is obtained, and tumor staging is performed. Staging aims to determine whether the tumor is confined to its site of origin or has spread elsewhere in the body. A bone marrow biopsy may also be performed if metastases are suspected.

Stage I - the tumor has not spread beyond its site of origin.

Stage II - the tumor has spread to adjacent structures, such as lymph nodes, but does not extend beyond the body's anatomic midline.

Stage III - The tumor extends beyond the midline; adjacent lymph nodes may or may not harbor tumor cells.

Stage IV - This stage is marked by widespread metastases, especially to the liver, lungs, and skeleton.

Stage IV-S (Special) - Although this form of neuroblastoma tends to be widespread at the time of diagnosis, it almost always responds well to treatment.

In a small number of cases, the neuroblastoma regresses spontaneously without the need for surgical intervention. In most cases, however, surgery, radiation, and chemotherapy are the major modalities for the treatment of neuroblastoma. Autologous bone marrow transplantation following high-dose chemotherapy may have benefits in children with advanced-stage tumors.

Prognosis

Most cases of Stage I, II, and IV-S neuroblastoma respond well to therapy; most other cases do not. The overall 5-year survival rate for neuroblastoma is around 60%; however, different stages of the disease have significantly different survival rates. Two key factors critical in determining prognosis are the patient's age at disease onset and the stage of the tumor. Other cytogenetic factors come into play as well.

1) Age - Patients younger than 1-year-old have a far better prognosis than all other age groups.

2) Stage - Patients with Stage I and IV-S tumors have an especially good prognosis, with a 5-year survival rate of over 95%. Patients with Stage III or IV tumors tend to have poor outcomes (and only a 10% survival rate at 5 years).

3) Ploidy of tumor cells - This refers to DNA copy number. Diploid or nearly diploid tumors tend to carry a poor prognosis. In contrast, hyperdiploid and triploid tumors often carry a better prognosis.

4) Chromosomal deletions. Deletions of the short arm of chromosome 1 occur in up to 80% of diploid tumors. Loss of this chromosomal segment is believed to result in the loss of tumor suppressor genes. Deletions of the long arm of chromosome 14 occur in as many as 50% of neuroblastomas and are often associated with aggressive tumors.

5) Amplification of the N-myc oncogene is another poor prognostic sign and is often associated with chromosome 1p deletions. Some tumors contain as many as 300 copies of the N-myc gene, which promotes cell division. As a rule of thumb, the higher the N-myc amplification number, the poorer the prognosis.

One marker associated with a favorable prognosis is high levels of the nerve growth factor (NGF) receptor protein Trk A. Patients with high Trk A levels almost never have tumors with N-myc gene amplification. These patients often experience regression of their tumors and generally have high survival rates.

Essay 80: Neurofibromatosis / Von Recklinghausen Disease

Joseph Merrick (1862-1890), known throughout England as the Elephant Man may have had the most severe case of neurofibromatosis (NF) documented in the annals of medical literature. In the 19th century, nothing was known about the genetic basis of NF or of any other disease for that matter. In light of his disfiguring condition, the traveling show Merrick worked for dubbed him the Elephant Man, and the popular press quickly followed suit. In 1882, a German physician named Von Recklinghausen published a monograph describing the appearance and symptoms of patients with NF; hence, in the older literature, NF is often called Von Recklinghausen disease.

NF is characterized by nerve sheath tumors called neurofibromas, which arise from the Schwann cells that myelinate peripheral nerve axons. These neurofibromas may arise on the skin surface or around internal nerve fibers anywhere in the body. Other signs of NF include patches of skin discoloration called cafe au lait spots; Lisch nodules in the iris of the eye; macrocephaly (enlarged head); scoliosis and other skeletal malformations, and a higher incidence of certain cancers, especially an adrenal medullary tumor known as pheochromocytoma.

Epidemiology

NF is believed to affect 100,000 people in the U.S., making it among the most common hereditary neurological diseases. Estimates of its incidence are 1 in 3,000 to 4,000 newborns worldwide. Although NF is an autosomal dominant disease, nearly 50% of cases appear sporadically. Medical geneticists think germ-line mutations in the NF-1 gene cause sporadic cases on chromosome 17 during meiosis.

Occasionally, cases of NF may be relatively mild or even go undiagnosed due to a phenomenon called variable expressivity of the abnormal NF-1 gene. In other words, although the role of NF-1 in causing neurofibromatosis is solidly established, the factors underlying the highly variable manifestations of this disease remain unknown.

This leads to an obvious question: What does the NF-1 gene normally do? NF-1 is thought to act as a tumor suppressor gene. The protein it encodes, called neurofibromin, inhibits cell proliferation in certain cell types, especially glial cells, an umbrella term for oligodendrocytes in the central nervous system and Schwann cells in the peripheral nervous system.

As with other diseases arising from tumor suppressor gene mutations, such as Von Hippel-Lindau and retinoblastoma, the tumors in NF arise from a specific cell type; other cell types tend to remain unaffected even though every cell in the person's body contains at least one defective NF-1 gene.

An Ongoing Debate

Several other questions remain unanswered about Merrick's condition. Why was Joseph Merrick's case so severe? And did Merrick even have NF? No one knows for sure. Although Merrick died in 1890, some of his remains, including hair samples, were preserved and subjected to DNA analysis. Unfortunately, a century had elapsed, and the test results were inconclusive. Some have proposed that Merrick's disease was actually a severe form of Proteus syndrome, an even rarer disorder linked to a tumor suppressor gene on chromosome 10 called PTEN. Unlike NF-1, however, PTEN mutations are found in a wide variety of cancers (brain, lung, prostate), none of which Merrick had. The debate continues.

Essay 81: Parkinson's Disease and the 1918 Flu Pandemic

Parkinson's disease (PD) is a degenerative motor disorder resulting from the progressive loss of dopaminergic neurons in an area of the brain known as the basal ganglia, in particular, two clusters of cells called the substantia nigra. The manifestations of PD appear when 80% or more of the dopamine-producing neurons have been destroyed. Symptoms of PD include difficulty initiating movement (bradykinesia), a shuffling gait, the classic pill-rolling hand tremor, a blank facial expression, and muscle rigidity; and in 10-15% of cases, the onset of dementia late in the course of the disease.

Nearly a century ago, in the wake of the 1918 influenza pandemic, which left as many as 50 million people dead worldwide, physicians reported numerous cases of a Parkinson-like condition described as post-encephalitic parkinsonism, also referred to as Von Economo's encephalopathy. Patients developed debilitating paralysis, sometimes years after suffering a bout of influenza. Whether infection with a particular strain of influenza virus was the culprit in the development of PD symptoms years later remains a mystery, as very few cases of post-encephalitic Parkinsonism have been diagnosed since 1926.

The enigmatic influenza virus

Although many scientists are convinced that the genotype of the influenza virus responsible for the 1918 pandemic was H1N1, the mechanism underlying this virus's virulence remains unclear. Part of the reason for this uncertainty stems from the unique nature of the influenza virus itself. Unlike most other viruses, influenza contains a segmented genome comprising eight distinct RNA sequences. The virus not only undergoes genetic point mutations, known as antigenic drift, over the course of a flu season but also a phenomenon known as antigenic shift.

Antigenic shift occurs when two strains of influenza viruses infect the same cell and then proceed to mix and match entire segments of their genomes. This genome-swapping ability leads to radically altered (and occasionally deadly) viral progeny, as occurred in later flu epidemics in 1957 and 1968. Incidentally, although several hundred thousand Americans died in those epidemics, there was no spike in the number of cases of postencephalitic Parkinsonism afterward.

Another factor complicating the picture is the ability of influenza to hide in hosts other than humans, especially in pigs, chickens, and ducks. All of these animals are capable of spreading the virus to humans; in fact, the H1N1 strain most closely resembles an avian influenza virus. Whether the survivors of the 1918 pandemic who went on to develop postencephalitic Parkinsonism were infected with a unique variant of H1N1 (or perhaps with multiple strains of influenza) will probably never be known for certain.

In addition to a viral etiology, several other theories have been proposed to account for the pathogenesis of PD. As yet, no consensus has emerged as to a specific cause or causes. In contrast to the etiology, the epidemiology of PD is well characterized: approximately 1% of Americans over age 65 will develop Parkinson's. The disease shows no predilection for any particular ethnic group, and both genders are affected equally. Although PD can strike younger people, they represent roughly 10% of the estimated 3 million cases of PD in the United States.

Since 90% of people with PD develop symptomatic disease after age 65, many researchers believe long-term exposure to environmental toxins is the major risk factor for developing PD.

Some contend that people who live in rural areas and drink well water are at a higher risk of developing PD. Other researchers focus on exposure to the chemical MPTP, which induces Parkinsonian symptoms in primates.

Proponents of MPTP exposure point to a high incidence of PD (or a similar disease called PSP) among Chamorro natives of the island of Guam, whose diet formerly included fruit bats. Some species of fruit bats are believed to contain high levels of MPTP or similar molecules in their tissues. Yet other researchers believe the alpha-synuclein gene on chromosome 4 is involved in PD, citing the example of a family with a rare autosomal dominant form of the disease. Nonetheless, most cases of PD occur sporadically, even among identical twins.

The treatment regimen of PD patients consists of the dopamine precursor Levodopa (L-dopa can cross the blood-brain barrier where it is converted to dopamine); anticholinergic medications to balance out the relative excess of acetylcholine in the basal ganglia; and several adjunct medications, including dopamine receptor agonists like bromocriptine (to mimic the action of dopamine), as well as monoamine oxidase and COMT (catecholamine O-methyltransferase) inhibitors to slow the degradation rate of dopamine throughout the brain. Efforts to replace dopaminergic neurons with stem cell transplants have met with mixed results. Evidence suggests that younger patients tend to benefit more than elderly ones from cell transplant therapy.

Essay 82: Patau Syndrome

Trisomy 13, also called Patau syndrome, results from three copies of chromosome 13 instead of the normal two copies. This disorder is relatively infrequent, occurring in approximately 1 in 10,000 live births. As with other human chromosomal disorders, most cases of Trisomy 13 are thought to arise from sporadic non-disjunction events during oogenesis. Advanced maternal age is associated with most cases of chromosomal trisomies. Rarely, chromosomal translocations can give rise to Patau syndrome. Even rarer phenomena include partial trisomy of chromosome 13 and mosaicism, in which only some cells contain an extra copy of the chromosome.

Many congenital abnormalities are associated with Trisomy 13. The most prominent anatomic anomalies include cleft lip or palate; microcephaly (small head); small, sometimes fused eyes; low set ears; micrognathia (small lower jaw); a variety of cardiac defects; umbilical and/or inguinal hernias; syndactyly (fused fingers); polydactyly (extra fingers or toes); other skeletal abnormalities; and cryptorchidism (undescended testes). Mental retardation and seizures are especially severe in babies with Patau syndrome.

The reason chromosomal trisomies lead to such pronounced physical abnormalities is not completely understood. One possibility is that the presence of an extra chromosome leads to a higher rate of aberrant cell divisions (sometimes termed mitotic catastrophe). Defective cell division, in turn, results in the formation of fewer functional cells during embryogenesis and fetal life. A shortage of healthy cells ultimately translates into aberrant cell migration, stunted growth, organ system malformations, and a relatively short life span.

Another factor thought to contribute to congenital anomalies is the phenomenon of gene dosage. Human cells contain two copies of most genes (the obvious exception being a single copy of the X and Y chromosomes in males). A third copy of most genes probably leads to excess production of the proteins encoded by those genes. In Trisomy 21, for example, a third copy of the APP gene results in brain pathology resembling Alzheimer's disease in most Down syndrome patients who live past age 40. Similar processes are no doubt at work in Patau syndrome.

Compared to Down syndrome, babies born with Trisomy 13 have a much lower survival rate, with an estimated 80% mortality in the first month of life alone. Treatment of Patau syndrome largely consists of supportive care. Women who have given birth to a baby with Patau syndrome have a small but elevated risk of a chromosomal trisomy in subsequent pregnancies. Trisomy 13 (as well as other chromosomal abnormalities) can be detected prenatally by karyotype analysis following amniocentesis or chorionic villus sampling.

Essay 83: The Philadelphia Chromosome and Leukemia

The so-called Philadelphia chromosome is an abnormal chromosomal structure present in most cases of chronic myelogenous leukemia (CML) and, to a lesser extent, in other types of leukemia, most notably acute lymphocytic leukemia (ALL). The origin of the term Philadelphia chromosome simply reflects the fact that a research team first identified this abnormality in a blood sample from a leukemia patient in Philadelphia, PA.

CML vs. Acute Leukemias

Unlike acute forms of leukemia, in which the cardinal manifestations (fever, weight loss, and fatigue) appear abruptly, patients with CML may be asymptomatic for months or even years prior to diagnosis. Even after diagnosis, many patients with chronic forms of leukemia survive 5-10 years or longer. In approximately 90% of CML patients, an abnormal structure known as the Philadelphia chromosome results from an event known as chromosomal translocation.

Basically, a piece of one chromosome breaks off and attaches itself to a non-homologous chromosome. In CML, a chromosomal translocation occurs between chromosomes 9 and 22. Geneticists abbreviate this event t9:22. The result of this t9:22 translocation is the juxtaposition of the c-abl tyrosine kinase gene on chromosome 9 with the bcr (breakpoint cluster region) on chromosome 22. The product is an abnormal fusion gene that encodes a constitutively active tyrosine kinase called BCR-ABL.

The BCR-ABL protein is a perpetually active version of a cell proliferation signal. The normal versions of these intracellular tyrosine kinases are generally inactivated unless the cell is responding to growth factors or cytokines that promote cell proliferation. The BCR-ABL protein is a different story. It acts like a broken record, continuously telling the cell to copy its DNA and undergo cell division. Consequently, lymphocytes that undergo the t9:22 translocation begin to proliferate out of control.

New Treatment Options: Tyrosine Kinase Inhibitors

Tyrosine kinase inhibitors represent a promising treatment option in CML and other cancers. The first successful drug in this class was Imatinib, better known as Gleevec, which inactivates the BCR-ABL protein, significantly extending the survival of patients with CML. Gleevec also treats a relatively rare cancer known as gastrointestinal stromal tumor. A similar drug called Lapatinib (Tykerb) has shown promise against certain forms of breast cancer.

Scientists have discovered that many cancers overexpress growth factor receptors and even secrete their own growth factors. Once these proteins bind to their respective cell surface receptors, they trigger similar cytosolic signal cascades.

In other words, similar messenger proteins transmit a signal from outside the cancer cell to various transcription factors and genes within the cell's nucleus that activate mitosis (cell proliferation). Tyrosine kinase inhibitors break this chain of communication, impairing (or ideally halting) the cancer cell's ability to proliferate.

Essay 84: Pituitary Tumors

The pituitary gland occupies a bony structure called the sella turcica, located at the base of the skull underneath the optic chiasm. The optic chiasm marks the site behind the eyes where axons from each optic nerve intersect en route to a part of the brain called the thalamus. The pituitary consists of an anterior region, where most of its hormones are produced, and a posterior part directly innervated by neurons in the hypothalamus. Most tumors in the pituitary gland involve the anterior portion, sometimes called the adenohypophysis. This article will cover the main types of pituitary tumors as well as the corresponding treatment options.

Non-functional Adenomas

Approximately 20% of pituitary tumors are non-functional adenomas, meaning they do not secrete detectable levels of any hormone. In the older literature, they are referred to as null cell adenomas or chromophobe tumors (due to faint staining with histological dyes). The vast majority of these tumors are not malignant, and small tumors (microadenomas < 1 centimeter in diameter) often remain asymptomatic. Tumors that grow beyond 1 cm in diameter are a different story. Their location all but guarantees several characteristic symptoms: frequent headaches, worsening visual disturbances, especially the loss of peripheral vision, and episodes of nausea and vomiting.

People exhibiting these red-flag symptoms of an intracranial mass should undergo diagnostic imaging studies, including a head CT and MRI. If a distinct mass lesion is spotted in the pituitary, the definitive treatment is transsphenoidal surgery. In this procedure, the surgeon inserts a flexible cable with a miniature camera into the patient's nasal passage, makes an incision in the sphenoid bone, and removes as much of the pituitary tumor as possible. Sporadic pituitary tumors may recur following excision in a small number of cases. Recurrences are more common in rare cases of pituitary carcinoma or in an uncommon genetic disorder called Multiple Endocrine Neoplasia Type I.

Prolactinomas and other functional adenomas

These tumors constitute around 75% of pituitary adenomas. These tumors generally arise from a single cell type in the pituitary and, with the exception of plurihormonal adenomas, secrete one particular hormone.

Scientists estimate that 20-30% of these adenomas secrete prolactin (Prl), which suppresses menstruation and can induce milk production even in non-pregnant women. 15% produce the gonadotrophins FSH and LH, which often disrupt the menstrual cycle as well. Another 15% release excess adrenocorticotropic hormone (ACTH), which manifests as an endocrine disorder called Cushing's disease. Approximately 5% produce excess growth hormone (hGH), which leads to acromegaly in adults and gigantism in children. The remaining 1% secrete thyroid stimulating hormone (TSH), resulting in an overactive thyroid gland.

Surgery remains the preferred treatment for patients with an enlarging or symptomatic pituitary adenoma. Patients undergo annual blood tests for 5-10 years after surgery to detect elevated hormone levels. A sustained rise in Prl, GH, TSH, etc., strongly suggests a recurrence of the tumor.

Compared to surgery, medical therapy for pituitary tumors is relatively limited. Bromocriptine is an option for patients with prolactinomas and GH-secreting adenomas who decline surgery or for whom surgery would be extremely risky.

Bromocriptine works by activating dopamine receptors on the prolactinoma cells, suppressing prolactin release. The drug also decreases GH release from somatotroph cells, which originate from the same cellular precursors as lactotrophs during embryonic development. Rare cases of hyperthyroidism caused by an inoperable TSH-secreting adenoma can usually be controlled with propylthiouracil (PTU) or other anti-thyroid drugs.

Essay 85: Pneumonia - CAP vs. Nosocomial Infections

Pneumonia refers to an infection of the lower respiratory tract. Although many classification schemes exist, e.g., bacterial/viral/fungal pneumonia, most clinicians find it useful to group pneumonia into two broad categories: community-acquired pneumonia (CAP) and nosocomial pneumonia.

CAP

Community-acquired pneumonia simply refers to any case of pneumonia contracted outside of a hospital. Most cases of CAP that come to medical attention are of bacterial origin. Although virtually any bacteria can cause pneumonia, organisms such as Streptococcus pneumonia (pneumococcus), Haemophilus influenzae, Moraxella catarrhalis, and Klebsiella species are the usual suspects. Factors such as a person's age, smoking status, and comorbid diseases influence individual susceptibility to bacterial pneumonia.

In any event, the classic presentation of CAP is fever and chills, fatigue, chest pain, and a cough productive of sputum developing over the course of hours to days. Bacterial pneumonia is usually treated empirically with broad-spectrum antibiotics such as fluoroquinolones and cephalosporins.

In contrast to bacterial pneumonia, viral pneumonia tends to produce less pronounced symptoms: a non-productive cough with or without fever persisting for several weeks. Symptoms are seldom severe enough to require hospitalization. For this reason, viral pneumonia is sometimes referred to as "walking pneumonia." Aside from oseltamivir (Tamiflu) for influenza, which is often a setup for pneumonia in the elderly and immune-compromised, no specific antiviral drugs are indicated for the treatment of viral pneumonia. Fortunately, most cases of viral pneumonia are self-limited and require little treatment beyond supportive care.

Atypical pneumonia presents much like viral pneumonia. The most common culprits include Mycoplasma and Chlamydia psittaci, a.k.a. parrot fever, which infected parakeets and chickens can spread. Although sputum cultures are sometimes taken, atypical pneumonia is generally treated empirically with azithromycin or doxycycline.

Nosocomial Pneumonia

This category of pneumonia is the opposite of CAP. Any case of pneumonia acquired during a hospital stay is referred to as nosocomial pneumonia. Hospital-acquired infections tend to be more resistant to antibiotics compared to CAP organisms. This is due to the spread of multidrug resistance plasmids among bacteria in hospitals, especially in critical care units, where a significant number of patients often have life-threatening infections.

Leading bacterial culprits in nosocomial pneumonia include Pseudomonas aeruginosa, which can colonize indwelling catheters and central lines, and Legionella pneumophila, a bacterium capable of growing inside air conditioning units and humidifiers, especially during periods of disuse. Suppose such equipment is used without being cleaned appropriately. In that case, Legionella organisms can be sprayed throughout a room, a unit, or even an entire building in the case of a contaminated central air conditioning system. Nosocomial infections often necessitate therapy with multiple broad-spectrum antibiotics.

Pneumonia due to Opportunistic Infections

Patients with HIV, cancer, and certain respiratory diseases (COPD, cystic fibrosis) are vulnerable to a variety of infections that would never gain a foothold in people with an intact immune system. Chief among these is Pneumocystis jirovecii (formerly known as P. carinii), an organism variously classified as a protozoan or a fungus, depending on which source one consults. The antibiotic Trimethoprim-sulfamethoxazole (Bactrim) is given as prophylaxis against Pneumocystis pneumonia in these patient populations.

Immune-compromised people are also susceptible to mycobacterial infections, including tuberculosis (new onset or reactivated) and mycobacterium avium intracellulare. The latter organism often exhibits multidrug resistance and may be difficult, if not impossible, to eradicate in an immune-compromised host.

Essay 86: Portal Hypertension

The term portal hypertension refers to any pathological process obstructing blood flow from the portal vein, through the liver parenchyma, and into the hepatic vein en route to the right side of the heart. Most cases of portal hypertension are caused by cirrhosis or scarring of the liver following chronic injury. In the U.S., the leading causes of cirrhosis are alcohol abuse and viral hepatitis (Hepatitis B or C). Less often, liver damage results from genetic disorders, including hemochromatosis, marked by iron overload; Wilson's disease, a rare disease resulting in copper overload; and autoimmune hepatitis.

In much of the developing world, the leading culprits include the Hepatitis B virus, especially in East Asia; mosquito-borne viruses like yellow fever, prevalent in tropical Africa, India, and South America; helminthic infections, especially liver flukes; and last but not least, protozoan infections such as visceral leishmaniasis, sometimes called kala-azar, or black water fever.

Signs and Symptoms

Obstruction of venous blood flow through the portal vein produces a constellation of abnormal findings. In mild to moderate portal hypertension, the body attempts to compensate by means of collateral venous circulation, in effect shunting blood away from the damaged liver. These collateral veins take the form of esophageal varices and hemorrhoids, both of which are prone to bleeding. Some patients develop clusters of spidery veins around the umbilicus, a condition known as caput medusa, or Medusa's head, owing to the tortuous appearance of these veins.

As portal hypertension worsens, the failing liver produces insufficient amounts of certain blood proteins, particularly albumin. This normally abundant protein acts as a sponge in the veins and capillaries, maintaining an adequate plasma volume. In hypoalbuminemia, however, fluid progressively seeps through the walls of these blood vessels and accumulates in the abdominal cavity, a condition known as ascites. Patients with severe ascites are at risk for developing SBP (spontaneous bacterial peritonitis) and dying of septic shock.

Treatment Options

Liver damage can be halted and sometimes reversed before the cirrhotic stage. People with alcohol dependence, for example, can quit drinking. People with hemochromatosis can undergo phlebotomy at monthly intervals to control their iron stores.

Viral hepatitis can be treated, or at least contained, with interferons. Parasitic infections can be eradicated, especially if diagnosed in their early stages. However, the treatment options dwindle once extensive liver damage occurs, giving way to cirrhosis.

Patients with ascites are often treated with a procedure called paracentesis, followed by intravenous albumin replacement. Paracentesis is a sophisticated term for puncturing a patient's abdomen with a large needle and removing the ascites fluid - sometimes several liters of it. These patients also receive antibiotics as prophylaxis against SBP.

Patients with end-stage liver disease sometimes undergo a procedure called TIPS, or Transjugular Intrahepatic Portosystemic Shunt. Basically, a tube is inserted through the jugular vein and maneuvered through the right side of the heart, past the inferior vena cava, and into the hepatic circulation, where it diverts blood from the portal vein into the hepatic vein, bypassing the liver in the process. The TIPS procedure relieves portal hypertension but only temporarily. Over the course of weeks to months, these patients accumulate toxic levels of urea and ammonia in their blood, culminating in hepatic encephalopathy, a terminal condition marked by delirium, coma, and eventually death. In this situation, the only potentially curative treatment remains a liver transplant.

Essay 87: Causes of Right Lower Quadrant Pain

As with acute or sudden onset pain anywhere in the body, the first priority is to rule out a life-threatening source. In the case of acute right lower quadrant pain, clinicians should assume it is appendicitis until proven otherwise. Fortunately, other telltale signs and symptoms tend to accompany appendicitis, including fever, nausea, vomiting, abdominal guarding (the muscle layer of the abdomen tenses involuntarily), and rebound tenderness (the person experiences worsening pain when shaken slightly or when pressure is removed from his/her abdomen). Appendicitis most commonly occurs in people under age 30, but cases have been reported at all ages.

In addition to infectious etiologies, discussing all possible causes of abdominal pain would take up several volumes. Rather than launch into a lengthy discourse, discussing the major categories of the right lower quadrant (RLQ) pain seems more sensible. As one might anticipate, sources of RLQ pain range from relatively minor to potentially fatal. Minor causes are, by their nature, self-limited. As a general rule, they seldom last for more than 24 to 48 hours. Examples include indigestion, food poisoning, traveler's diarrhea, and most cases of viral gastroenteritis. Symptoms tend to resolve quickly once the offending agent(s) are removed from the body.

In contrast, certain events should be treated as red flags: constipation lasting 5 or more days, diarrhea lasting longer than 48 hours (or more than 24 hours in infants), bleeding from the rectum (other than hemorrhoids), fever, severe nausea/vomiting, and swelling, redness, or tenderness over any part of the abdomen. Other than diarrhea and bleeding, the other signs strongly point to an intestinal obstruction.

Small bowel obstruction (SBO) may cause pain anywhere in the abdomen, including RLQ pain. One mnemonic to help remember the main causes of SBO is SHAVITTS.

S = Stone, meaning gallstone ileus (lack of intestinal peristalsis). An impacted gallstone can obstruct the flow of bile or pancreatic juice, which in turn triggers intestinal inflammation. In this situation, peristalsis comes to a halt.

H = Herniation - these include femoral, inguinal, and obturator hernias, as well as a condition called Meckel's diverticulum (an outpouching of all intestinal layers forming a hernia sac). Although the latter condition is mostly diagnosed in patients age 2 or younger, cases can occur at any age.

A = Adhesions - loops of the bowel may stick together due to scar tissue formation after surgery or penetrating abdominal trauma.

V = Volvulus - The bowel flops on top of the mesentery, strangulating its own blood supply. This bizarre event may happen to people born with intestinal malrotation or other anomalies or for no apparent reason at all. People with poorly controlled diabetes have been known to delay seeking medical attention, as they sometimes experience this life-threatening event as a dull ache.

I = Intussusception - A proximal portion of the bowel telescopes into a distal segment. This condition is more common in children but can happen at any age. Often, it is triggered by an underlying tumor or a Meckel's diverticulum.

T = Tumor - Approximately 1% of gastrointestinal tract tumors occur in the small intestine. Most of the remainder arise in the large intestine. In either location, a large tumor can cause bowel obstruction.

T = Tapeworm - Although this parasite is uncommon in the United States, these worms can infect people who consume raw or undercooked beef, pork, or fish. Tapeworms can reach lengths of several meters before causing a person sufficient distress that they seek medical attention.

S = Strictures - these narrowed segments of the bowel are seen most often in patients with IBD (Inflammatory Bowel Disease), particularly Crohn's Disease. Radiographic images taken after a barium swallow show a so-called string sign indicative of intestinal strictures.

In contrast to SBO, the causes of large bowel obstruction are few, most notably tumors, discussed previously; a genetic condition called Hirschsprung's disease; and toxic megacolon. Hirschsprung's disease occurs when a person is born lacking nerve ganglia in the distal segment of the colon. This problem is generally detected in the first weeks of life when an infant is unable to pass stool. It is corrected by surgically resecting the aganglionic piece of the colon.

Just as its name suggests, a toxic megacolon occurs when a dead or dying colon becomes massively enlarged and inevitably ruptures its toxic contents into the abdominal cavity, usually resulting in death from septic shock. Toxic megacolon can arise as a complication of volvulus, IBD, or a parasitic infection known as Chagas' Disease, caused by the trypanosome T. cruzi. Chagas disease is common in tropical countries like Brazil but is seldom seen in the U.S.

Essay 88: Sickle Cell Anemia

Sickle cell anemia (SCA) is a widespread hemoglobin synthesis disorder, sometimes called hemoglobinopathy. As with other hemoglobinopathies, SCA is an autosomal recessive disease, meaning a person must have two defective genes present for the full-blown disease to occur. People who carry a single sickle cell gene are said to have the sickle cell trait. The children of two carriers have a 25% chance of inheriting SCA.

Epidemiology

In the U.S., SCA occurs most often among African-Americans, approximately 10% of whom are carriers of the sickle cell gene. This prevalence results in SCA in approximately 0.2% of live births in the African-American population or about 1 in 500. In parts of Africa, over 30% of the population has the sickle cell trait. A key reason for this high carrier rate is its survival value in areas where malaria is endemic. Before the discovery of quinine and other anti-malarial drugs, very few European colonists in tropical Africa or Southeast Asia survived longer than a few years.

Although malaria is most prevalent in tropical Africa and India, the malarial protozoan Plasmodium probably originated in Southeast Asia. The most likely scenario is that chickens were first domesticated from jungle fowl in Vietnam, and some of these birds carried malaria. As chicken farms spread westward, mosquitoes followed, spreading malaria to humans who often lived alongside chickens and other livestock. The main evidence supporting this theory is that today, most quinine-resistant malaria strains can be found in Thailand rather than India or Africa.

Disease Manifestations

From a clinical standpoint, the main concern is the so-called sickle crisis, which occurs when someone with SCA is exposed to low oxygen environments, cold temperatures or is suffering from an infection. Most of his/her red blood cells lose their flexible biconcave shape and become rigid crescents, which easily obstruct capillaries. In addition to causing excruciating pain, these sickle crises can lead to potentially fatal complications, namely heart attacks and strokes.

In addition to the high risk of heart attack and stroke, most people with SCA lose the function of their spleens by adolescence. This is believed to result from small numbers of sickled red cells progressively damaging the spleen over time. In medical terms, this loss of the spleen, functional asplenia, or autosplenectomy, leaves SCA patients with a compromised immune system. They are especially susceptible to infections by encapsulated bacteria, including salmonella, streptococcus pneumoniae, meningococcus, and Haemophilus influenzae.

Treatments

In childhood, the main treatment regimen for SCA involves minimizing sickle crises, thereby lowering the risk of juvenile heart attack or stroke (a truly catastrophic event). In addition to the vaccines recommended for all children, SCA patients should receive a salmonella vaccine, a childhood Pneumovax against strep pneumonia, a meningococcal vaccine, and arguably, doses of the Hib vaccine against H. influenzae into adolescence.

As for treating SCA on the molecular level, a few therapies seem promising. One drug, hydroxyurea, appears to decrease the expression of the mutant beta-globin gene even as it increases the expression of fetal hemoglobin genes, whose proteins can substitute for the hemoglobin beta chain. Beyond hydroxyurea, other treatments for SCA are still in the preliminary stage. One idea is to remove stem cells from the patient's bone marrow, insert healthy beta-globin genes, and then transplant these genetically engineered cells back into the patient.

Bone marrow transplants have met with occasional success; however, the procedure carries several major risks, especially graft versus host disease (GVHD). Also, finding compatible bone marrow donors in the African-American community is often challenging.

Essay 89: Systemic Lupus Erythematosus

Systemic Lupus Erythematosus (SLE) is an autoimmune disease affecting an estimated 1.5 million people in the United States. 90% of patients diagnosed with SLE are female. Although no convincing explanation accounts for SLE's cause(s) or the skewed gender ratio, current theories tend to invoke a combination of ill-defined environmental and genetic factors. Because the signs and symptoms of SLE can vary markedly from patient to patient, the diagnosis may be delayed or missed altogether. This article will focus on the criteria physicians use to diagnose SLE and explain why treating this disease remains such a clinical challenge.

The classic presentation of SLE is a woman of childbearing age with a butterfly-shaped facial rash accompanied by unexplained fevers, weight loss, fatigue, and arthritis. Blood work typically reveals anemia (low red blood cell count), a decreased white blood cell count, and the presence of abnormal proteins called antinuclear antibodies (ANA), which will be discussed in the next section.

Other red flags include skin rashes that worsen with exposure to sunlight, oral ulcers, visual disturbances, unexplained seizures, and protein in the urine of a patient with no prior history of kidney disease. Unfortunately, in many cases of SLE, the patient experiences only vague symptoms, especially muscle aches and joint pain, which mimic rheumatoid arthritis (RA) as well as other connective tissue diseases.

The 1997 Updated Criteria for Classifying SLE

To bring some order to the diagnostic chaos, in 1982, the American College of Rheumatology compiled a checklist to assist physicians in diagnosing SLE. These criteria were last updated in 1997. Suppose a patient exhibits four or more of the following eleven findings (sequentially or in combination) at any time after seeking medical attention. In that case, a diagnosis of SLE can be made with 85% certainty.

(1) Malar rash – the butterfly-shaped facial rash

(2) Discoid rash – unlike a malar rash, this type can leave permanent scars.

(3) Photosensitivity to UV light

(4) Oral ulcers

(5) Arthritis – tenderness, pain, or swelling without bony erosions in 2 or more peripheral joints

(6) Serositis refers to inflammation of the pericardial sac surrounding the heart or pleural membrane around the lungs. Pericarditis is confirmed with an EKG; pleuritis may or may not be detectable with imaging studies.

(7) Renal disorder – persistent protein in the urine or debris called cellular casts on microscopic analysis. Untreated lupus nephritis can progress to end-stage renal disease, necessitating dialysis or a kidney transplant.

(8) Neurologic disorder includes seizures or psychiatric disturbances in a patient with no prior history of these conditions and who is not on any drug/s that could induce these findings.

(9) Hematologic disorder – characterized by low red blood cells (anemia), low white blood cells (leukopenia), or low platelet counts (thrombocytopenia).

(10) Immunologic disorder – this category includes three distinct auto-antibodies: Anti-Sm (Smith antigen), Anti-double stranded DNA and antiphospholipid antibodies. In addition, a false positive test for Treponema pallidum, the bacteria that causes syphilis, falls under this category.

(11) Positive ANA – Antinuclear antibodies occur in almost all patients with SLE. ANA (as well as other auto-antibodies) bind to normal proteins, forming so-called immune complexes. These complexes travel throughout the blood circulation and eventually settle in small blood vessels, most notably in the kidneys, eyes, and brain. Other blood proteins called complement enzymes attach themselves to the immune complexes, which in turn recruit white blood cells to the scene. The result is continuous low-grade inflammation and chronic tissue damage. Almost all SLE patients have positive ANA titers; however, this finding is common in other autoimmune diseases like RA. In the language of clinicians, a positive ANA result is sensitive but not specific to SLE.

What happens next?

As with other autoimmune diseases, there is no cure for SLE. The treatment goals are, first, to halt acute flare-ups of the disease and second, to prevent end-stage organ damage caused by chronic inflammation. Glucocorticoids, such as high-dose prednisone, remain the treatment of choice to combat acute flare-ups of SLE. However, physicians try to avoid long-term steroid use due to its adverse effects: diabetes, hypertension, increased risk of infection, and osteoporosis. Treatments for skin rashes include retinoids, dapsone, quinacrine, and routine use of sunscreen with SPF > 15.

Maintenance therapy for SLE generally consists of anti-malarial drugs and chemotherapy agents, including azathioprine, cyclophosphamide, and chlorambucil. Stronger immune suppressant drugs (normally reserved for organ transplant patients) like cyclosporine and mycophenolate have been used if all else fails. Experimental treatments include plasmapheresis, intended to remove immune complexes from the circulation, and infusions of the monoclonal antibody Rituximab, which inhibits B-lymphocyte activity and perhaps reduces auto-antibody production. With current medical treatment options, the 5-year survival rate of SLE patients is approximately 90%, with a 20-year survival rate of 75%.

Essay 90: Sports Injuries

The term sports injury encompasses bone fractures, which often require emergency medical attention; sprains, which include stretched or torn ligaments; and strains, which refer to stretched tendons, also known as pulled muscles. In each of these situations, inflammation begins immediately after the injury occurs. The hallmarks of inflammation include pain, swelling, warmth, and redness at the site of injury.

From a medical standpoint, the acute inflammatory response is an elaborate response designed to contain tissue damage. Small blood vessels dilate to increase blood flow to the area, accounting for redness and warmth. Fluid leaks out of damaged blood vessels and lymphatics, contributing to swelling. Damaged endothelial cells, white blood cells, and platelets release mediators, including histamine, bradykinin, prostaglandins, and serotonin. These mediators enhance the sensation of pain. They also recruit more inflammatory blood cells to the scene, exacerbating the pain and swelling in the process.

Given this situation, applying heat to an acute sports injury will only make the pain and swelling worse. Therefore, as a general rule, ice packs should be used for the first 48 to 72 hours following a sports injury. Ice packs should not be applied continuously, however. Applying ice for 15 minutes on and then 15 minutes off is a safe guideline. A well-known mnemonic for treating an acute sports injury outside of a hospital setting is RICE, which stands for Rest, Ice, Compression, and Elevation.

The benefits of rest are largely self-explanatory, but for the sake of those religiously devoted to working out, avoid exercise or strenuous physical activity during this time. Ice is explained in the preceding paragraph. Compression refers to an ace bandage or any form of wrap designed to minimize swelling. Elevation means keeping the injured limb above heart level. That way, gravity will cause blood to flow away from the site of injury rather than pool there.

After 48 hours or so, acute inflammation tends to subside. (If this does not occur, a bone fracture or other serious trauma must be ruled out). Heat can be used at this point, with appropriate safety measures, of course. Hot surfaces should always be wrapped in cloth and never applied directly to the skin.

Electric heating pads should only be used at low to moderate settings and should always be disconnected before a person falls asleep. Children, older people, and people with diabetes are vulnerable to suffering burns from electric heating pads. People with long-standing diabetes often have peripheral nerve damage and may sustain a second-degree burn without feeling significant pain or discomfort.

Essay 91: Stages of Chronic Kidney Disease

Chronic kidney disease, abbreviated CKD, affects well over 1 million people in the U.S. today, although a significant fraction of them are asymptomatic or remain undiagnosed. Common causes of CKD include long-standing diabetes mellitus, poorly controlled hypertension, glomerulonephritis, and genetic disorders like Adult Polycystic Kidney Disease.

Regardless of its cause, the hallmark of CKD is a progressive decline in the volume of blood plasma filtered by the kidneys per unit time, a value known as the glomerular filtration rate (GFR). Normal adult GFR is considered to be over 90 milliliters per minute, abbreviated mL/min. The severity of CKD corresponds to the decline in GFR. In the current scheme, CKD is divided into five stages, with each stage being more severe than the previous one.

Stage 1: GFR > 90 mL/min.

In Stage 1 of CKD, there are no outward signs or symptoms. GFR is basically normal at > 90 mL/min., but evidence of kidney damage is present, generally in the form of microalbuminuria (small amounts of the blood protein albumin) detected during urinalysis. Most cases of Stage 1 CKD occur in the context of long-standing diabetes or hypertension; hence, a diagnosis of CKD is often anticipated, if not entirely predictable. No specific treatment exists other than to correct or manage the underlying condition contributing to CKD.

Stage 2: GFR = 60 – 90 mL/min.

As with Stage 1, people with Stage 2 CKD are usually asymptomatic. Treatment efforts are geared toward halting, or at least slowing, further decline in kidney function. Many nephrologists are convinced that a class of blood pressure medications called ACEIs (Angiotensin Converting Enzyme Inhibitors) helps preserve kidney function in patients with Stage 2 CKD. Although trials will no doubt proceed for the foreseeable future, most clinicians see no downside to starting an ACE inhibitor in this patient population.

Stage 3: GFR = 30 - 60 mL/min.

At this stage, the trouble becomes more serious. Patients with Stage 3 CKD often develop high blood pressure and anemia. High blood pressure probably results from a combination of fluid retention and excessive renin/angiotensin hormonal system activation. Anemia, marked by a loss of red blood cells, is most likely due to declining erythropoietin production by the failing kidneys. Erythropoietin normally boosts red blood cell production in the bone marrow. Another concern is the imbalance between serum levels of calcium and phosphate. Partly because of decreased sensitivity to PTH from the parathyroid glands, the failing kidneys reabsorb too little calcium and too much phosphate. If the serum calcium and phosphate product exceeds 55 (mg/dL) 2, the patient is at high risk for developing kidney stones, an extremely painful condition.

Patients with Stage 3 CKD are encouraged to minimize or eliminate phosphorus-rich foods, such as soy products and cheeses, from their diet. If the dietary modification does not control phosphate levels, the drug Sevelamer can be taken before each meal, although this drug is usually reserved for patients with Stage 4 or 5 CKD. Sevelamer is a resin that binds to phosphate in the intestines and prevents its absorption into the bloodstream. Vitamin D supplements are also prescribed to promote calcium absorption from the diet as well as calcium reuptake in the kidneys.

Stage 4: GFR = 15 - 30 mL/min.

This stage marks the cut-off for initiating kidney dialysis. In some cases, ambulatory patients may opt for peritoneal dialysis, which can be performed at home and requires far less equipment than the alternative procedure, hemodialysis. Hemodialysis is performed at specialized dialysis centers or in a hospital setting twice to thrice weekly. In the U.S., virtually everyone who needs dialysis is covered under Medicare.

Patients who receive dialysis too infrequently develop uremia, a condition marked by excessive levels of urea and uric acid in the blood that cannot be eliminated. Some patients with uremia develop a skin condition (uremic snow) caused by the sweat glands' increased excretion of uric acid. The concentrated uric acid forms a white film on the skin, producing intense, unbearable itching.

Stage 5: GFR < 15 mL/min.

In the medical literature, stages 4 and 5 of CKD are collectively known as end-stage renal disease (ESRD). Patients with poorly managed ESRD develop severe uremia, leading to metabolic acidosis (because blood pH drops), electrolyte imbalances, especially hyperkalemia (high potassium), and death due to coma or cardiac arrhythmias. Aside from routine dialysis, a kidney transplant remains the only definitive treatment for ESRD.

Essay 92: Staphylococcus – the Ultimate Superbug

Staphylococcus is a gram-positive bacterium that occurs almost ubiquitously; most people carry staph organisms on their skin surfaces and nasal passages. Intact skin and mucosa are generally effective barriers to staph infections. Unfortunately, the converse is also true, with staph being the chief culprit in wound infections and abscesses. This article will focus on the two main groups of staphylococci and the infections they cause.

Staphylococcus aureus (henceforth abbreviated S. aureus) got its name from its ability to form gold-colored colonies. S. aureus can also tolerate sodium chloride concentrations as high as 15%, or about five times the salt content of seawater. This bacterium is often the cause of wound infections, the formation of abscesses, or walled-off pockets of infection. Unlike most other bacteria, S. aureus possesses an enzyme called coagulase, which causes human blood to clot, effectively shielding large numbers of these bacteria from destruction by white blood cells.

S. aureus is second only to streptococci as a leading cause of infective endocarditis and subacute bacterial endocarditis, abbreviated IE and SBE, respectively. S. aureus is thought to be the culprit in 30% of cases of native valve endocarditis. Common symptoms of endocarditis include fever, a new onset heart murmur, splinter hemorrhages in the nail beds, joint and muscle pain, and anemia. Untreated IE is generally fatal within 6 weeks; even with treatment, the mortality rate is around 30%. In contrast, untreated SBE may take up to a year to be fatal and has an excellent prognosis if treated.

In addition to abscesses and endocarditis, S. aureus also causes an array of other infections, including food poisoning, Toxic Shock Syndrome, Staphylococcal scalded skin syndrome, and osteomyelitis (bone infection). Immune-compromised people are also susceptible to pneumonia, meningitis, and other systemic infections caused by S. aureus.

Treatment of Staphylococcus aureus infections

Most S. aureus abscesses require surgical debridement and drainage. In addition to wound care, a course of broad-spectrum antibiotics is warranted. First-line agents include cephalosporins, especially cefazolin; piperacillin/tazobactam (Zosyn); and nafcillin.

Over the past several years, however, multiple drug-resistant strains of S. aureus have emerged in hospitals, collectively known as MRSA (methicillin-resistant staph. Aureus). The one reliable treatment for MRSA is a glycopeptide antibiotic called Vancomycin. Many physicians and microbiologists are convinced that, in the not-too-distant future, strains of Vancomycin-resistant staphylococci (VRSA) will become a serious problem in hospitals across the United States.

Coagulase Negative Staphylococci

Clinicians sometimes refer to all other strains of staphylococci as coagulase-negative staph, or CNS for short. CNS is a leading cause of nosocomial or hospital-acquired infections. Staph saprophyticus mainly causes urinary tract infections, predominantly in patients with indwelling bladder catheters. These UTIs generally respond well to antibiotic therapy.

Another CNS, Staph epidermidis, can cause serious infections due to its tendency to colonize heart valves, especially prosthetic ones. A potentially fatal complication of S. epidermidis infection is prosthetic heart valve rupture. This form of CNS is also thought to cause up to 6% of native valve endocarditis infections as well.

Essay 93: Tumors of the CNS - Diagnosis and Treatment

With the advent of advanced imaging techniques, including CT, MRI, and PET scans, brain cancer diagnosis can be made more readily than ever before. In comparison, the treatment options for brain cancer, especially for malignant astrocytomas, remain limited, and the majority of cases still carry a poor prognosis. After exposure to innumerable books, movies, and documentaries about brain tumors, most people are aware of the "red flag" symptoms: excruciating headaches, double vision or blurred vision, projectile vomiting, new onset seizures, and the progressive loss of sensory, motor, and cognitive functions.

Brain cancer can be divided into two general types: primary, meaning the tumor originated in the brain, or metastatic, meaning the tumor originated elsewhere in the body and then spread to the brain. Let's discuss the second category first.

A variety of cancers can spread to the brain, namely lung cancer, breast cancer, choriocarcinoma (a form of testicular cancer), as well as melanoma (a dreaded skin cancer that can spread almost anywhere in the body). In some cases of choriocarcinoma as well as osteogenic sarcoma, chemotherapy drugs or radiation can eradicate the metastases. In most other cancers, however, brain metastases indicate a poor prognosis, with death ensuing in 6-12 months. These patients are often treated with X-ray therapy, which temporarily alleviates the symptoms of these metastatic cells. However, radiation therapy seldom induces remission, and the metastatic cancer cells generally reappear weeks to months later.

Once primary brain cancer is diagnosed, a staging biopsy is performed. This procedure aims to classify the tumor based on its histology (cell of origin) and grade (aggressiveness of the tumor cells). The cells of origin are most often glial cells, in particular astrocytes and oligodendrocytes. CNS lymphoma is often grouped with primary brain cancers but is seen mainly in immune-compromised patients (AIDS, cancer, organ transplant recipients).

Other intracranial tumors are not considered brain cancer in the strictest sense of the term. These include meningiomas, ependymomas, and pituitary adenomas.

Most of these tumors are not malignant and can often be resected surgically; nevertheless, they can press against adjacent brain structures and produce the classic symptoms of headache, vomiting, seizures, and visual disturbances.

In terms of grade, high-grade tumors display features that portend a poor outcome. These include an undifferentiated appearance, high mitotic figures (large numbers of actively dividing tumor cells), and angiogenesis, the formation of new blood vessels that supply the growing tumor with nutrients and oxygen. In contrast, low-grade tumors tend to progress more slowly but often transform into high-grade tumors over time. This is thought to occur as less aggressive tumor cells die off and are replaced by more malignant cells that are resistant to chemotherapy drugs.

As noted above, the treatment options for brain cancer still lag far behind the diagnostic modalities. In fact, the 5-year survival rate for astrocytomas has remained almost unchanged over the past 50 years. Chemotherapy agents include the nitrosoureas BCNU and CCNU, which are derivatives of mustard gas, cisplatin, and etoposide.

More recently, cytokines like Interleukin-12 have been tried in the hopes of triggering an immune response against the tumor.

In addition, the monoclonal antibody drug bevacizumab (Avastin) starves the tumor of nutrients by inhibiting blood vessel formation and extending life by 3-6 months. Non-drug therapies include surgery (mainly to debulk the tumor mass) and targeted radiation, especially the gamma knife.

Future therapies will likely include tyrosine kinase inhibitors (TKIs). Many astrocytomas overexpress the EGF (Epidermal Growth Factor) receptor, which acts as an out-of-control tyrosine kinase, perpetually driving tumor cell proliferation. Some of these TKI agents, like Gleevec and Tykerb, are already used to treat CML leukemia and breast cancer. An alternative approach involves monoclonal antibodies conjugated with radioisotopes or bacterial toxins. Ideally, these agents would selectively target tumor cells while sparing healthy brain tissue.

Essay 94: Von Hippel-Lindau Disease

Von Hippel-Lindau disease (VHL) is a relatively rare genetic disorder affecting an estimated 10,000 people in the U.S. The inheritance pattern of VHL is autosomal dominant. This means that anyone born with the disease gene has a 50% chance of passing it on to his or her children. VHL is marked by the formation of capillary tumors (hemangioblastomas) in the central nervous system, most often in the cerebellum and retina. Another disease manifestation is the growth of cysts in the liver, kidneys, or pancreas.

In 10% of people with VHL, one or more of the hemangioblastomas release erythropoietin, a hormone that stimulates the production of red blood cells in the bone marrow. The result is a condition called polycythemia vera, an abnormally high number of red blood cells. A notable symptom of polycythemia is intense itching, especially after bathing. Occasionally, the excessive number of red blood cells leads to the formation of blood clots, a potentially fatal complication.

Individuals with VHL also have an increased risk of developing renal cell carcinoma as well as pheochromocytoma, a chromaffin cell tumor of the adrenal medulla that can release high levels of epinephrine and norepinephrine. This can result in sudden surges in blood pressure, leading to syncope (fainting spells) as well as life-threatening complications, especially stroke.

The genetic basis of the disease involves a mutation in the VHL tumor suppressor gene located on chromosome 3. The mutant VHL gene encodes an abnormal protein (pVHL) that interferes with the elongation step of messenger RNA synthesis. The mutant protein is believed to compete with a normal protein called elongin A, which forms a ternary complex along with elongins B and C. The absence of the normal elongation factor complex allows for the production of abnormally long messenger RNA molecules.

Over time, as large numbers of mutant mRNAs are produced, the odds of translating oncogenic proteins increase dramatically. Ultimately, this mechanism contributes to the high incidence of tumors in people with VHL. As to why such relatively uncommon tumor types arise in VHL patients (renal cell carcinoma and pheochromocytoma), no one knows for sure.

Treatment of VHL involves surgical resection of symptomatic hemangioblastomas as well as nephrectomy and/or adrenalectomy (surgical removal of a kidney or adrenal gland) in cases of renal cell carcinoma or pheochromocytoma, respectively. Patients with bilateral tumors of the kidneys can undergo partial nephrectomy, but they tend to have a poorer prognosis. On the other hand, bilateral removal of the adrenal glands can be treated with oral hydrocortisone to replace cortisol and Fludrocortisone (Florinef) to replace the mineralocorticoid aldosterone.

Pharmacology

Essay 95: Anti-Arrhythmic Drugs A Double-Edged Sword

Amiodarone is the generic name of an anti-arrhythmic drug marketed as Cordarone or Pacerone. Amiodarone is classified as a Class III anti-arrhythmic, meaning it mainly acts as a potassium channel blocker. This prolongs the plateau phase of the cardiac myocyte action potential, increasing calcium influx and extending the time before the next action potential begins. In addition to potassium channel blockade, amiodarone exhibits the actions of Class I, II, and IV antiarrhythmics – sodium channel, beta-adrenergic receptor, and calcium channel blockade, respectively.

Although it has gained widespread acceptance as an all-purpose antiarrhythmic, amiodarone has some serious side effects. These include an overactive or underactive thyroid gland, pulmonary fibrosis, yellow or brown discoloration of the eye sclera, skin discoloration, an extremely long elimination half-life ranging from 2 weeks to 3 months, and numerous drug-drug interactions, most notably with warfarin and digoxin. (Some old-time cardiologists are distressed by this interaction. Digoxin is not merely a drug; for old-school cardiologists, it is a religion). On top of all that, amiodarone occasionally triggers the very arrhythmias it was meant to suppress, a problem seen with other antiarrhythmics as well as certain antipsychotic drugs and inhaled anesthetic agents.

Procainamide is the generic name of an anti-arrhythmic medication marketed as Procan or Pronestyl. Procainamide is categorized as a Class Ia anti-arrhythmic drug. Class Ia drugs block voltage-gated sodium channels in the Purkinje cells of the heart's conduction system. Blocking the influx of sodium ions slows the depolarization phase of a cardiac myocyte's action potential. These drugs also lengthen the repolarization phase in cardiac pacemaker cells, slowing a person's heart rate in the process. Although procainamide is still prescribed orally, it is more often used as an emergency drug in patients with sudden cardiac arrest.

As with amiodarone, procainamide has several notable side effects. First, if the drug is given intravenously, it can cause hypotension or a sudden drop in blood pressure. This effect occurs in approximately 5% of patients who receive procainamide by the IV route. Hypotension is believed to result from the blockade of peripheral nerve ganglia located along the aorta.

Second, procainamide triggers the formation of antinuclear antibodies (ANA) in approximately 50% of people who take the drug long-term. Half of these patients, in turn, go on to develop a lupus-like syndrome characterized by arthritis, abdominal pain, fever, and rash. Pericarditis and pleural effusion have also been reported in association with drug-induced lupus.

Drug-induced lupus is thought to develop mainly in patients who are slow metabolizers of procainamide. Fortunately, this adverse reaction tends to subside once procainamide is discontinued. Because of this potential side effect, however, procainamide is contraindicated in patients with SLE (systemic lupus erythematosus) or a positive serum ANA titer. The antihypertensive drug hydralazine is another cardiovascular medication known to induce a lupus-like syndrome.

Third, an NIH study called CAST (Cardiac Arrhythmia Suppression Trial) found that in a small percentage of patients, anti-arrhythmic drugs paradoxically induced cardiac arrhythmias by prolonging part of the cardiac conduction cycle known as the QT interval.

Prolongation of the QT interval raises the risk that dysfunctional cardiac conduction cells will fire inappropriately, inducing the same electrical disturbances the drugs were meant to control in the first place.

In the worst-case scenario, this side effect can result in a potentially fatal outcome: ventricular fibrillation, or V-fib.

For this reason, procainamide is contraindicated in patients with second-degree AV block not controlled by an artificial pacemaker or those with a history of an arrhythmia called torsades de pointes. Procainamide also contains an FDA black box warning, recommending that its use be limited to patients with life-threatening ventricular arrhythmias.

Several other side effects have been reported with procainamide, but these tend to occur in less than 1% of patients. They include severe allergic reactions (angioedema), agranulocytosis (a severe drop in white blood cell count), aplastic anemia (total bone marrow failure), hemolytic anemia (probably due to autoantibody-mediated destruction of red blood cells); inflammation of the myocardium; liver failure; pancreatitis, and a neurologic disorder called demyelinating polyneuropathy.

Returning to the question posed in the title of this essay, yes, certain anti-arrhythmic drugs are a double-edged sword. They are life savers in some contexts but, on occasion, may be fatal. Much like fire and other modern technologies, we have come to depend on, these drugs have become a necessary evil in the realm of medicine.

Essay 96: Antibiotics

In a general sense, an antibiotic is defined as any natural or synthetic chemical capable of neutralizing a pathogen, either by inhibiting its growth or directly killing the organism. When scientists and physicians refer to antibiotics, however, they virtually always mean drugs used to combat bacterial infections instead of antiviral, antifungal, or anti-parasitic medications. With that in mind, this article will focus on the major categories of antibiotics in use today.

Penicillin

Natural penicillin was isolated by Alexander Fleming in 1928. After observing that bacteria were unable to form colonies near a species of bread mold called Penicillium, Fleming deduced that the mold was producing microbicidal chemicals. Penicillins are beta-lactam antibiotics. They work by inhibiting a bacterial enzyme called transpeptidase, which catalyzes the formation of peptidoglycan, the backbone of bacterial cell walls. Penicillins cover many gram-positive bacteria, which have thick peptidoglycan layers in their cell walls, and certain gram-negative organisms, whose cell walls contain much less peptidoglycan than their gram-positive counterparts.

Some strains of staphylococcus and strep bacteria remain susceptible to penicillins, but many have acquired resistance, largely because of years of indiscriminate use. Penicillin still remains the treatment of choice for syphilis, an STD that remains prevalent in the U.S. and many other countries. Other conditions treated by penicillin include strep throat, middle ear infections, and endocarditis.

Because penicillin can bind to the serum protein albumin, this antibiotic triggers more allergic reactions than other classes of antibiotics. Approximately 2% of the population is thought to exhibit some form of allergy to penicillin. Penicillin allergies range from hives and rashes to, in rare cases, life-threatening anaphylaxis, marked by sudden swelling of the face and trachea. Death may result unless the person is treated promptly with epinephrine.

Cephalosporins

Cephalosporins are also beta-lactam antibiotics, first discovered in the 1940s in an Italian sewage system. This may sound strange until one considers antibiotics' role in nature: microbes synthesize them to wage war on other microbes.

Cephalosporins are used to treat a variety of infections ranging from cellulitis to pneumonia and meningitis. The most recently developed cephalosporins treat pseudomonas, a gram-negative organism that can cause life-threatening infections in immune-compromised patients.

As of 2013, a 5th generation cephalosporin called Ceftaroline (Teflaro) came on the U.S. market. Ceftaroline is the only beta lactam capable of killing MRSA. This antibiotic is a valuable addition to the infectious disease arsenal, as the rise of Vancomycin resistant Staph aureus is only a matter of time.

Carbapenems

These are broad-spectrum beta-lactam antibiotics. They cover nearly all bacteria with the notable exception of MRSA (Methicillin Resistant Staphylococcus Aureus). Carbapenems can be given only by the intravenous (IV) route; as such, they are used almost exclusively in the hospital setting.

Vancomycin

This glycopeptide antibiotic produced by soil bacteria in Borneo was first used clinically in the late 1950s. Vancomycin's mechanism of action is similar to penicillin; it irreversibly inhibits an early step in bacterial cell wall synthesis.

Vancomycin is one of the few antibiotics capable of killing MRSA, a multidrug-resistant strain of Staphylococcus. Because Vancomycin covers virtually all gram-positive bacteria, it remains one of the most important antibiotics used in hospitals today.

Macrolides

This class of antibiotics works by inhibiting protein synthesis in bacteria by binding to the 50S ribosomal subunit. Macrolides include erythromycin, clarithromycin, and azithromycin. These drugs mainly cover gram-negative bacteria such as H. pylori (linked to stomach ulcers) and H. influenzae, a frequent culprit in sinusitis and pneumonia.

Quinolones

These antibiotics include Cipro (ciprofloxacin), Levaquin (levofloxacin), and Avelox (Moxifloxacin). Quinolones work by inhibiting the bacterial enzyme DNA gyrase, which in turn interferes with bacterial DNA replication and gene expression. Cipro is mainly used to treat urinary tract infections, while Levaquin and Avelox are used to treat sinusitis and pneumonia.

Sulfonamides

The major drug in this category is Bactrim, generically known as TMP-SMZ (Trimethoprim-Sulfamethoxazole) or Co-trimoxazole. Its mechanism of action is to block bacterial DNA synthesis by inhibiting the key enzyme involved in folate metabolism known as DHFR (dihydrofolate reductase). Bactrim is used to treat urinary tract infections and occasionally for traveler's diarrhea. Patients with an allergy to sulfa drugs should avoid Bactrim.

Tetracyclines

This class of broad-spectrum antibiotics works by inhibiting bacterial protein synthesis. Newer tetracyclines like doxycycline effectively treat Chlamydia, Lyme disease, and atypical pneumonia. Newer drugs in this class include minocycline and tigecycline. Tetracycline itself was overused for so many years that most bacteria acquired resistance to it. Today, it is mainly used to treat acne.

Aminoglycosides

These antibiotics include gentamicin, tobramycin, amikacin, and others. Aminoglycosides cover aerobic gram-negative bacteria exclusively. Their mechanism of action is to bind to the 30S ribosomal subunit, effectively halting bacterial protein synthesis. Since they are not absorbed orally, aminoglycosides must be given intravenously. As such, they are used to treat serious gram-negative infections in hospitalized patients.

Anaerobic coverage – Clindamycin and Metronidazole

These two antibiotics cover anaerobic bacteria above and below the diaphragm, respectively. Clindamycin (Cleocin) has a similar mechanism of action to macrolides, inhibiting bacterial protein synthesis by blocking the 50S ribosomal subunit. Metronidazole (Flagyl) is effective against anaerobic intestinal bacteria such as Clostridium difficile as well as protozoa such as Giardia and other amoebae.

Antimycobacterial agents

These include isoniazid, rifampin, pyrazinamide, ethambutol, dapsone, and clofazamine. The first four agents are mainly used to treat tuberculosis (TB), whereas the last two are almost exclusively used to treat leprosy. TB and leprosy are caused by mycobacteria – slow-growing bacteria whose cell walls resemble those of fungi. This thick cell wall makes mycobacteria impervious to most antibiotics; mycobacterial infections must be treated for months or even years before therapy can be stopped.

Miscellaneous Antibiotics

These run the gamut from narrow-use agents like nitrofurantoin (used exclusively to treat urinary tract infections) to broad-spectrum drugs like Zyvox (linezolid), daptomycin (a lipopeptide), and the streptogramin Synercid (quinupristin/dalfopristin), which are used to treat pneumonia, sepsis, and other life-threatening infections. The newest class of antibiotics, called gramicidins, targets bacterial ion channels, which deprive the bacteria of energy by halting ATP production.

Essay 97: Caffeine

The alkaloid 1,3,7-trimethylxanthine, more commonly known as caffeine, is found in many natural sources, including coffee, cocoa beans, and tea leaves. Although caffeine has been used as a CNS stimulant for centuries, its exact mechanism of action is not completely understood. Caffeine mimics the actions of epinephrine (adrenaline) in that a high enough dose can raise heart rate and blood pressure. It also seems to stimulate gut motility and act as a diuretic. In the CNS, caffeine may act on certain adenosine receptors to promote alertness and enhance attention span.

Although caffeine may mask symptoms such as fatigue, no evidence exists to support the claim that caffeine can alleviate or shorten the withdrawal symptoms of alcohol intoxication, a.k.a. a hangover. This is true mainly because the human body requires a minimum number of hours to readjust after metabolizing a higher dose of ethanol than it is normally accustomed to handling. To illustrate this concept more fully requires a foray into the subjects of pharmacodynamics and pharmacokinetics.

Pharmacodynamics refers to any measurable effect of a drug on the human body. In the case of ethanol, these effects include an altered sense of judgment and mood changes at moderate doses, impaired motor coordination and drowsiness at higher doses, and, at toxic doses, respiratory depression and coma. An individual's response to alcohol is influenced by a host of factors, including gender, body weight, frequency of alcohol consumption, and a family history of alcoholism.

Pharmacokinetics focuses on what the human body does to a particular drug. Scientists are concerned with four parameters in pharmacokinetic studies, abbreviated as ADME: absorption, distribution, metabolism, and excretion. As will be explained, the metabolism of ethanol plays a central role in the length and severity of a hangover.

With rare exceptions, ethanol enters the body exclusively by the oral route. After entering the stomach, ethanol tends to be absorbed quickly across the gastric mucosa, especially in the absence of fatty food. Milk and other foods high in fat may delay ethanol absorption, but contrary to popular myth, they do not inhibit the absorption of alcoholic beverages.

The amphipathic nature (both lipid and water soluble) of ethanol allows it to cross cell membranes with relative ease.

Ethanol also achieves a large volume of distribution. This means that ethanol does not remain confined to the bloodstream; instead, it enters virtually every tissue in the body, especially the central nervous system (CNS).

Even as ethanol exerts its effects on the CNS, the liver begins to metabolize it, after which the kidneys excrete the metabolites. The kinetics, or rate, of hepatic metabolism, sets ethanol apart from most other toxins metabolized by the liver.

All other drugs, except aspirin and phenytoin, undergo metabolism by first-order kinetics in the liver. This means that for therapeutic doses of the drug, the liver ramps up the production of the appropriate enzymes to metabolize the drug in question.

Not so for ethanol.

This is one of a handful of chemicals metabolized by zero-order kinetics. Basically, this means whether a person has consumed a single alcoholic beverage or has just engaged in a drinking binge, the liver can metabolize no more than a specific amount of ethanol per unit of time, regardless of the concentration of ethanol present in the bloodstream. In other words, although the amount of ethanol an individual can metabolize varies from person to person, the liver's capacity to handle ethanol is already "maxed out" prior to the first drink.

As such, the greater the amount of ethanol the liver must handle, the longer symptoms of intoxication will persist and the more likely it is for withdrawal symptoms (the hangover phase) to occur. Although caffeine is metabolized in the liver, it does not stimulate increased production of CYP2E1 and alcohol dehydrogenase, the major enzymes involved in ethanol metabolism. Thus, with or without coffee, a hangover will last the same length of time.

Essay 98: Ergot Alkaloids

Ergot is a group of alkaloids produced by a fungus prevalent in northern Europe and other damp climates. The fungus, known as Claviceps purpurea, can contaminate certain grains, particularly rye, barley, and other grasses. Classically, victims of ergot poisoning suffer three major side effects: burning pain in the extremities, miscarriages, and CNS effects, especially hallucinations, and seizures.

Intense vasoconstriction in the arms and legs

In the older medical literature, this was referred to as St. Anthony's fire. People suffering from ergot poisoning often complain of an intense burning sensation in their extremities. Today, scientists understand that muscle and other tissues deprived of blood flow and oxygen burn glucose anaerobically, generating large amounts of lactic acid. Lactic acidosis's hallmark symptoms include muscle fatigue and a painful burning sensation in the oxygen-starved tissues. In extreme cases, untreated vasoconstriction can result in gangrene of the hands or lower legs.

Ergot alkaloids induce miscarriage.

As with extremity pain, this side effect is also caused by the interruption of blood flow. When the uterus is starved for blood, it tends to undergo powerful muscle contractions, sometimes to the point of continuous muscle spasms (called tetani). As a result, the placenta separates from the uterine wall, and the woman goes into preterm labor.

Hallucinations and seizures

The basis for these side effects stems from a combination of vasoconstriction and the chemical similarity of ergot to the hallucinogenic drug lysergic acid diethylamide (LSD). The exact basis of LSD-induced hallucinations remains obscure. Activation of NMDA and serotonin receptors resulting in altered firing patterns of serotonergic neurons is believed to be involved. Apparently, ergot alkaloids can cross the blood-brain barrier and trigger hallucinations by similar mechanisms. CNS hypoxia secondary to decreased blood flow may account for seizures.

The symptoms of ergot poisoning account for much of the witch hysteria that gripped Europe during the late Middle Ages.

The most prevalent episodes of witch hunts occurred, not coincidentally, in those parts of Europe that grew large amounts of rye and barley, for example, Germany, Scandinavia, and the British Isles.

The climate pattern from the 14th through the early 19th century (sometimes referred to as the Little Ice Age) favored the damp, chilly conditions that promoted ergot growth. In the minds of a society living in a pre-scientific era, the only explanation for a mysterious epidemic of spontaneous miscarriages and hallucinations was witchcraft. As late as 1692, ergot poisoning may have fueled the Salem Witchcraft Trials in New England.

Today, ergot alkaloids are mainly used to stop bleeding in women with postpartum hemorrhage. They were also commonly used to treat migraine headaches but have fallen into disuse with the development of newer drugs like sumatriptan and other serotonin receptor antagonists.

Essay 99: Incredible Insulin

Insulin is a peptide hormone produced by the beta cells in the islets of Langerhans, which comprise the endocrine pancreas. Unlike most other hormones, which directly or indirectly elevate blood glucose, insulin is one of the few hormones that lower blood sugar levels in addition to promoting the anabolism of amino acids into proteins and free fatty acids into storage as triacylglycerol chains.

The main trigger for insulin release is a rise in blood glucose, which occurs minutes to hours after a meal (depending on whether or not the food contains a significant amount of simple sugar). The beta cells that manufacture insulin sense blood sugar levels via a receptor called GLUT-2 (Glucose Transporter-2).

When blood glucose exceeds a certain concentration (approx. 120 mg/dL in euglycemic people), the GLUT-2 receptors transport glucose inside the beta cell, leading to a burst of ATP production. This spike in ATP leads to the closure of potassium ($K+$) channels and opening of calcium ($Ca2+$) channels, resulting in the depolarization of the beta cell membrane and exocytosis of insulin in a manner analogous to neurotransmitter release from a neuron following an action potential.

As with all endocrine glands, the beta cells release insulin directly into the bloodstream, which acts on numerous tissues, especially the liver, skeletal muscle, and adipose (fat) tissue. Insulin remains active for 15-30 minutes, after which most is degraded by cells in the liver and kidneys.

Insulin's mechanism of action was not understood until recently, and elucidating the effects of this hormone remains an intense area of research. Like other peptide hormones, insulin cannot cross cell membranes, so it must exert its effects by binding to a cell surface receptor. The insulin receptor is known as a receptor tyrosine kinase (RTK). It is a homodimeric protein, meaning it consists of two identical subunits.

When insulin binds to its receptor, the two subunits move close together, and several tyrosine residues on each subunit become phosphorylated. At this point, the RTK is activated. In its activated state, the RTK adds phosphate groups to nearby proteins, including one called Insulin Receptor Substrate (IRS). Somehow, the phosphorylated form of IRS induces the movement of glucose transport channels (e.g., GLUT-3, GLUT-4) to the cell surface.

This allows the liver hepatocytes, skeletal muscle cells, adipocytes, etc., to take up glucose from the extracellular fluid and utilize it for energy.

Insulin's effects on protein and fat metabolism are also anabolic, but the signal transduction pathways involved are less well understood. There is a time delay of several hours between insulin binding to a target cell and an increase in protein or lipid synthesis. This suggests that insulin's long-term effects involve alterations in gene transcription. Changes in gene expression may occur via several pathways, including PKA/CREB, Ras/MAP Kinase, PI3 Kinase, and others.

Deranged glucose metabolism, in the form of the disease diabetes mellitus (DM), is a leading cause of morbidity and mortality in the industrialized world today. The hallmark symptoms of diabetes include malaise, fatigue, polydipsia (excessive thirst), polyphagia (excessive hunger), and polyuria (excessive urination, sometimes in excess of two gallons of urine in a day).

The complications of untreated diabetes are well known: retinopathy leading to blindness, neuropathy, gangrene of the extremities, a general susceptibility to infections, poor wound healing, and kidney failure.

Traditionally, diabetes was classified as Type 1/Insulin dependent/Juvenile onset DM vs. Type 2/ Non-insulin-dependent/Adult onset DM. Although the usefulness of this scheme has been called into question, it remains helpful in understanding the onset and progression of most cases of diabetes.

Type 1 diabetes is believed to be an autoimmune disease in which the beta islet cells are destroyed, leading to an absolute insulin deficiency. Type 1 DM represents about 10% of all diabetes cases. Most cases occur sporadically, meaning they do not show a clear hereditary pattern. Most Type 1 DM patients come to medical attention by adolescence. For unclear reasons, the incidence of Type 1 DM is on the rise in many countries in the world.

A life-threatening complication seen mainly in people with Type 1 DM is a condition called Diabetic Ketoacidosis (DKA). In DKA, the body acts as though it were starving in the midst of plenty. Although blood glucose levels may exceed 500 mg/dL, the liver converts proteins into acetoacetate and B-hydroxybutyrate, collectively known as ketone bodies. This metabolic pathway normally kicks in during periods of prolonged starvation. Out-of-control production of ketone bodies overwhelms the body's ability to maintain blood pH at 7.4. The person's blood turns acidic, leading to nausea, vomiting, and sometimes seizures and coma. Untreated DKA is generally fatal.

The treatment for Type 1 DM is insulin injections, which have become more manageable thanks to the development of Lantus insulin, administered once a day. Future therapies will likely involve islet cell transplants and the creation of durable islet cells from stem cell progenitors.

In contrast to Type 1 DM, patients with Type 2 DM tend to be overweight adults who also suffer from high blood pressure and high cholesterol. Type 2 DM is not caused by insufficient insulin production; it results from end-organ insensitivity to insulin. Basically, the liver, muscles, and fat tissue ignore insulin and fail to take up blood glucose.

A life-threatening complication in Type 2 diabetics is called a non-ketotic hyperosmolar coma. NKHC occurs due to a combination of physiological stress and lack of access to water. Blood glucose levels may exceed 1000 mg/dL, and the person becomes progressively dehydrated until s/he slips into a coma.

Treatment of early-stage Type 2 DM consists of one or more oral medications. Over time, 25% or more of Type 2 diabetics require insulin injections. An especially effective drug in managing Type 2 DM is Metformin (Glucophage), which decreases glucose production in the liver. Other drugs, e.g., sulfonylureas, act as secretagogues that promote insulin release from the pancreas. Still, other drugs, such as Acarbose, block glucose uptake in the small intestine.

In the 1990s, a new class of diabetes drugs came on the market. These were the thiazolidinediones (TZDs) that act as insulin sensitizers in the liver and other tissues. Unfortunately, two of these drugs, Rezulin (troglitazone) and Avandia (rosiglitazone) were linked to liver failure and heart attacks, respectively. Ultimately, Rezulin was withdrawn from the market at the behest of the FDA. Avandia was withdrawn from the European market, and prescriptions for it have plummeted in the U.S. Another TZD called Actos (pioglitazone) remains on the market; however, retrospective research suggests a link (never proven) between Actos and an increased risk of bladder cancer. Predictably, the use of Actos has declined as well.

Even newer treatment options for Type 2 DM include the injectable drug Byetta (Exenatide) and the oral drug Januvia (Sitagliptin). These medications enhance the actions of insulin by promoting its release or slowing its degradation respectively. As with the TZDs, they are useful adjunct therapies only when insulin is present. Unlike the TZDs, neither drug produces life-threatening side effects. Hence, these medications have supplanted Actos in the treatment of Type 2 DM.

Since 2010, the newest class of oral diabetes drugs is the SGLT inhibitors. These drugs work by blocking glucose reabsorption in the kidneys via the sodium/glucose co-transporter channels in the proximal convoluted tubules. These include canagliflozin (Invokana), dapagliflozin (Farxiga) and empagliflozin (Jardiance). These medications increase sugar lost in the urine, lowering the amount of sugar circulating in the bloodstream. While this is helpful in the management of diabetes, adverse effects include dehydration as well as increased risk of urinary tract infections - because bacteria and yeast gravitate to body surfaces coated in sugar.

Essay 100: Melatonin

Melatonin is an indoleamine or chemical derived from the amino acid tryptophan. It is structurally related to the neurotransmitter serotonin but seems to exert somewhat different effects on the central nervous system. The main site of melatonin production is the pineal gland, a structure located on the brain's midline near the diencephalon, consisting of the thalamus, hypothalamus, epithalamus, and subthalamic nuclei.

The pineal gland receives signals from a group of neurons called the SCN (suprachiasmatic nucleus). Exposure to bright light (even for a few seconds) inhibits melatonin production, whereas prolonged exposure to dim light has the opposite effect. People who work the night shift for long periods of time or spend the winter in polar regions often develop chronically elevated melatonin levels. In response to a prolonged lack of sunlight, the body enters a quasi-hibernation state characterized by perpetual sleepiness, lethargy, and sometimes clinical depression.

Aside from extreme environments, melatonin levels normally rise and fall in synchrony with a person's sleep-wake cycle. Melatonin levels are lowest in the morning (especially after exposure to sunlight). They slowly rise during the day and peak around bedtime. After several hours of sleep, melatonin levels decline in preparation for a new circadian cycle.

The above, somewhat simplified melatonin cycle can be thrown off kilter by a host of external factors: stimulants like caffeine, a variety of medications, and the most familiar cause, jet lag. If a person's insomnia is caused by jet lag, there is a good chance melatonin can help. On the other hand, many forms of insomnia are precipitated by events largely unrelated to melatonin, especially psychological stress, anxiety disorders, major depression, or the long-term abuse of stimulants, especially drugs like amphetamines or cocaine. Hence, determining the factor(s) contributing to insomnia is instrumental in formulating a treatment plan for insomnia.

For straightforward cases of insomnia related to jet lag, over-the-counter melatonin tablets such as Melatonex may be taken according to the directions on the bottle. Melatonin supplements also come in soluble lozenge forms for children and people who cannot swallow tablets easily. Regardless of whether or not it induces sleep, children should take no more than 0.3 mg of melatonin in a 24-hour period. Adults may take up to 5 mg of melatonin per day. The first dose should cause drowsiness in 30-60 minutes. Another dose may be taken if the person does not fall asleep within 2 hours. If a second dose is not effective, the person's insomnia is probably due to a cause other than a disturbance of his or her sleep-wake cycle, and supplemental melatonin will probably do no good.

Essay 101: How Do Neurotoxins Work?

The term neurotoxic venom refers to any biological molecule that interferes with the normal function of a neuron, sometimes culminating in cell death. The animals most often associated with neurotoxic venom are snakes and spiders. As we shall see, however, many aquatic species also produce neurotoxins, including puffer fish, jellyfish, sea anemones, scorpion fish, cone shells, and octopods. This article will focus on the different mechanisms of action of neurotoxins and briefly discuss the treatments available for people bitten or stung by these creatures.

Cobra toxin

Cobras are one of the most well-known venomous snakes. The neurotoxic component of cobra venom paralyzes skeletal muscles by blocking the nicotinic acetylcholine receptor, much like the plant compound curare and certain anesthetic drugs called neuromuscular blockers. The diaphragm is a skeletal muscle containing nicotinic acetylcholine receptors activated in response to the neurotransmitter acetylcholine released by the phrenic nerves. Death can occur several hours after a cobra bite due to respiratory muscle paralysis.

The treatment for a cobra bite is elapid antivenin purified from horse serum. This preparation contains equine antibodies to the toxins present in cobra venom. Since these are not human-derived antibodies, repeated injections of antivenin will trigger an immunological response called an Arthus reaction, also known as serum sickness. For patients who have respiratory muscle paralysis, mechanical ventilation is necessary.

Black widow spiders

After a spider bite, this toxin travels through the bloodstream, where it enters cholinergic neurons and causes a massive burst of acetylcholine release. This neurotransmitter causes sustained contraction, or tetany, of respiratory and skeletal muscles followed by paralysis. Tissue necrosis also occurs at the bite site. Treatment consists of airway support (in the form of a mechanical ventilator) along with an injection of antibodies to black widow venom.

Pufferfish

In Japan, puffer fish, also known as fugu, is a risky delicacy. Chefs who prepare puffer fish must be specially licensed; still, approximately 50 Japanese die annually after eating puffer fish. The reason is that female puffer fish produce a chemical called tetrodotoxin (TTX) in their ovaries. Some TTX can also be found in the fish's liver and skin. TTX has no distinct taste or odor but acts as a powerful blocker of voltage-gated sodium channels, stopping the propagation of action potentials along neuronal axons. Death occurs over the course of hours due to respiratory paralysis.

Certain species of mussels produce saxitoxin, a chemical with the same mechanism of action as TTX. The poison from a single mussel is estimated to kill 50 people.

Blue Ringed Octopus

The blue-ringed octopus is found mainly in the Pacific Ocean and Australian coastal waters, producing TTX like the puffer fish. Since there is no specific antidote for TTX, a bite can be fatal in a matter of minutes.

Jellyfish and sea anemones

These sea creatures contain specialized stinging cells called nematocysts that inject venom in a manner similar to a harpoon. Compared to snake venom, jellyfish venom is less well characterized.

From numerous reports, however, scientists know that jellyfish venom can interfere with respiratory muscle function and cardiac conduction. Regarding danger to humans, the most notorious jellyfish include the box jelly and the Portuguese man-of-war. Unlike jellyfish, anemones are relatively stationary and present little danger to swimmers or scuba divers.

Cone shells

These mollusks produce a poison called omega conotoxin, which acts as a powerful calcium channel blocker. Within a few seconds of exposure to this toxin, neurons stop releasing neurotransmitters. The reason is that although the neuronal axons can still conduct action potentials (mediated by sodium ion influx and potassium ion efflux through voltage-gated ion channels), neurotransmitter release requires an influx of calcium ions into the axonal terminal. Once conotoxin reaches the phrenic nerves, they stop releasing acetylcholine; the diaphragm muscle stops contracting, and death from respiratory muscle paralysis occurs a few minutes later.

Essay 102: Courses to Take Prior to Pharmacy School

Author's note: this article reflects my personal experience in applying to pharmacy school. Keep in mind that PharmD programs in the United States have similar, but not identical, course requirements.

Most applicants to pharmacy schools have earned a bachelor's degree in the realm of the life sciences. Some have a background in chemistry, and a few come from other academic disciplines, such as engineering. A few make the transition to pharmacy school after only two years of undergraduate coursework. At any rate, regardless of your undergraduate major, it is a good idea to take certain courses in preparation for pharmacy school.

In the realm of chemistry, at least one semester of inorganic chemistry and one semester of organic with their respective labs are required by many, if not most, schools of pharmacy. As for physics, one to two semesters of this course and the lab are also good ideas. Although the main relationship of physics to the field of pharmacy is the realm of pharmaceutics (the formulation and production of medications), one benefit is that it will keep your math skills sharp. And it will make courses like pharmaceutical chemistry less of a nightmare.

Biology presents an array of options. From personal experience, biochemistry, and cell biology are two extremely helpful courses in terms of scoring high on the Pharmacy College Admission Test (PCAT) and doing well in pharmacy school. As is neurobiology, human anatomy, physiology, and immunology are also good choices. If possible, try to take an introductory pharmacology course as well.

As far as a mathematics requirement goes, at least one semester of calculus is a good recommendation. It will definitely improve your score on the math section of the PCAT. At many undergraduate schools, all science majors must take one to two semesters of calculus anyway. My personal theory is that if calculus were to remain optional, no one except math majors would take it at the college level. Look on the bright side - at least you won't have to take really difficult math courses like differential equations. Also, make sure to take a 3 credit course in statistics. This course is also often a prerequisite, even though the important points are reviewed during first-year pharmacy school courses.

Other prerequisite courses may include such diverse subjects as English composition, microeconomics, and public speaking. Taking an English composition course makes sense to the extent that honing your writing skills never hurts. Public speaking is also good, especially for anyone with a phobia of speaking in front of a group. Microeconomics introduces you to the basics of running your own business. Considering that the number of independent pharmacies in the U.S. continues to decline, however, the logic behind making this course a prerequisite for pharmacy school escapes me.

The Top 10 things I learned in Pharmacy School

1. Older people fall down. A lot.
2. Without the U.S. Department of Health, all humans would still live in caves and eat raw meat cut with stone knives.
3. Warfarin interacts with literally everything.
4. So does Amiodarone.
5. We granulate powders to help them flow better. That's half of the pharmaceutics right there.
6. Scrape the icing off the cupcake. Don't argue. Scrap it off. Just do it.
7. Prisoners in Kentucky like drinking bourbon. The alcoholic ones do, anyway.
8. They're all overmedicated.
9. They're not getting the help they so desperately need.
10. They're all on drugs. Every last one of them.

Essay 103: Quetiapine and the CDS Schedule

Quetiapine is the generic name of an antipsychotic drug marketed under the trade name Seroquel. This medication is one of the so-called atypical antipsychotics, a class of newer agents used to treat thought disorders, in particular schizophrenia. Although Seroquel's mechanism of action is not completely understood, it is believed to act at certain dopamine and serotonin receptors, alleviating hallucinations as well as combating the symptoms of depression that often accompany schizophrenia.

As with other antipsychotic medications, Seroquel has several predictable side effects ranging from drowsiness and blurred vision to muscle rigidity, cardiac arrhythmias, and, in a small number of patients, a condition called tardive dyskinesia (a potentially irreversible motor disorder associated with long term use of antipsychotic drugs). One unusual side effect observed during animal studies of Seroquel was the development of cataracts. Although this side effect rarely occurs in humans, anyone prescribed Seroquel on a long-term basis should be examined by an ophthalmologist every six months.

Given its potentially dangerous side effects, an obvious question arises: why is Seroquel not considered a Controlled Dangerous Substance (CDS) like morphine or fentanyl? The short answer is that, unlike opiates, amphetamines, benzodiazepines, or barbiturates, Seroquel is not a drug of abuse; therefore, the chances of anyone becoming addicted to it are virtually zero. For a more in-depth answer, it is necessary to explain the scheduling system as well as the rationale for assigning certain prescription drugs a CDS designation.

By the 1950s, the Federal government enacted legislation that outlawed certain drugs entirely (e.g., heroin and marijuana) and placed restrictions on dispensing many others, especially morphine and its derivatives (collectively termed opiates). Predictably, additional legislation led to the need for a schedule, or hierarchy, of controlled substances. Ultimately, it was left to the DEA (Drug Enforcement Agency) to devise a classification system taking into account the legal status of certain drugs and the abuse potential of others.

Schedule I drugs, which include marijuana, LSD, and heroin, have no approved medical use. Ownership or use of these drugs is illegal under federal law, with the exception of certain federally exempted projects conducted by the NIH or similar organizations.

Schedule II (C-II) drugs include opiates with high abuse potential, such as morphine, oxycontin, and methadone. The amphetamine derivatives methylphenidate (Ritalin) and dextroamphetamine (Adderall) are also classified as C-II drugs. In order to legally prescribe Schedule II-V medications, a physician must be registered with the DEA. In addition, no refills can be written for C-II drugs, and no more than a 90-day supply of a C-II drug may be dispensed at any one time. Furthermore, some states have placed a time limit of 60 days on filling prescriptions for C-II drugs.

Schedule III (C-III) covers barbiturates such as thiopental (sodium pentothal), anesthetic agents like ketamine as well as most anabolic steroids. Opiates like codeine and buprenorphine are also designated C-III drugs.

Schedule IV (C-IV) drugs have less abuse potential than C-II or C-III drugs. C-IV drugs are mostly benzodiazepines, such as the anesthetic midazolam (Versed), as well as commonly prescribed anxiolytics like diazepam (Valium) and sleep aids.

Schedule V (C-V) drugs have minimal abuse potential. One example is the anti-diarrheal agent diphenoxylate (Lomotil).

Some states designate all other prescription drugs as "Schedule VI"; however, this label carries no meaning above and beyond "by prescription only." In practice, the DEA focuses its enforcement efforts on Schedule I and II drugs, as they represent the vast majority of drugs sold illegally in the U.S.

Essay 104: Dr. Katz's CVA Cocktail / X-Glut-5

Dr. Katz's CVA Cocktail / X-Glut-5

I've included this section in the book because, for many years, I have harbored an intense interest in the field of stroke research, in particular neuroprotection. I have always wanted to publish this idea, although I've never been directly involved in neuroprotection research. Given my background in both medicine and research, however, I am convinced that CNS damage, especially in the aftermath of a stroke, must be treated with a cocktail of medications as opposed to therapy with thrombolytic drugs alone. What follows is a portion of my research proposal that I would like to share with my intellectual peers, and hopefully, my future colleagues.

I call it the Dr. Katz approach to acute stroke management. If you're a neuroscientist, neurologist, pharmacist, or medical student, I think you will find my ideas intriguing. And if any attending physician out there has the guts to try it, by all means, let me know the outcome.

A New Approach to Acute Stroke Management

A. Background CNS neurons are totally dependent on aerobic respiration for ATP production. Ischemia, defined as an interruption of blood flow, is the most common etiology of CNS hypoxia in humans. A disruption in blood flow for as little as 10 seconds can lead to harmful (albeit reversible) metabolic alterations in CNS cells. Ischemia lasting for minutes, as in the case of a transient ischemic attack (TIA) produces a hypoxic insult to the brain that is largely reversible assuming blood flow is restored. Longer periods of hypoxia invariably result in brain tissue infarction. In humans this manifests as CVA (cerebrovascular accident) a.k.a. stroke, one of the five leading causes of mortality in the United States today.[1] Over 85% of strokes are thromboembolic in nature; the remainder include hemorrhagic strokes, intracranial hemorrhage due to head trauma, and prolonged vasospasm. The pathogenesis of neuronal cell death following hypoxic insult can be thought of as a "three strikes" scenario

a) Strike One: Glutamate excitotoxicity – the inability of hypoxic neurons to maintain ionic gradients leads to membrane depolarization and excessive release of glutamate at glutamatergic synapses. This effect is exacerbated by the impaired ability of hypoxic astrocytes to remove glutamate from the extracellular fluid.

b) Strike Two: Excessive glutamate binds to ionotropic receptors, including NMDA receptors, which allows uncontrolled calcium influx into neurons and glial cells.

c) Strike Three: The many toxic effects of elevated cytosolic calcium include the persistent activation of the phosphatase calcineurin; its subsequent dephosphorylation of nitric oxide synthetase (NOS); uncontrolled production of nitric oxide (NO) and other free radicals; damage to the mitochondria and nuclear DNA; and ultimately, the activation of apoptotic pathways culminating in cell death. Clinically, stroke patients can receive thrombolytics, especially tissue plasminogen activator (tPA), for up to 180 minutes following the onset of stroke symptoms. A crucially important consideration is that thrombolytics are indicated exclusively for the treatment of thromboembolic strokes. They are contraindicated in hemorrhagic strokes or in patients with bleeding disorders or who are on chronic anticoagulant therapy. In order to distinguish a thromboembolic stroke from a hemorrhagic one, a head CT must be performed, further delaying the initiation of thrombolytic therapy.

B. My CVA Cocktail to the Rescue

A properly designed drug cocktail would provide distinct advantages over the current approach to acute stroke management in the following ways: It would a) extend the time window for administering thrombolytics; b) have no significant contraindications; c) provide neuroprotection in any stroke regardless of etiology; and d) combat each of the above mentioned threats to cell viability.

The problem, as I see it, is a certain reluctance among clinicians to use a drug cocktail due to concerns about unforeseen drug interactions and adverse side effects. Their motto remains, "What if it doesn't work, and I get sued?" Well, my motto has always been, "No guts, no glory!" With that in mind, here is my untested, but potentially promising, five drug CVA cocktail. The names of the drugs are secret for now in the off chance the combination becomes a billion dollar blockbuster drug. Nonetheless, anyone who bothers reading the references can figure out the combination quite easily. Don't forget, I own the patent. So don't get any ideas ;-P

a) A1 – This anesthetic drug is the only parenteral NMDA receptor antagonist already in clinical use. Drugs in this category would shield neurons from excess glutamate, potentially limiting its excitotoxic effects.

b) A2 (CNI) – This immunosuppressant drug exhibits neuroprotective effects in rodent models of CNS ischemia.[2] A2's benefits may involve multiple mechanisms. First, a complex of A2 and an immunophillin binding protein inhibits the cytosolic phosphatase calcineurin, thereby decreasing the dephosphorylation of iNOS (an isoform of the enzyme known as inducible NOS), which then minimizes production of NO, albeit in macrophages more so than in neurons or glial cells. Second, A2 is known to inhibit IL-2 gene expression. This decreases T-cell infiltration into the hypoxic brain region, which in turn limits CNS inflammation and edema. Third, A2, probably in conjunction with FKBP12, induces de novo RNA synthesis in astrocytes within two hours of incubation. The identities of the target genes are unknown but may involve the upregulation of anti-apoptotic genes such as bcl-2 and bcl-XL.[3] Another study suggests the drug may downregulate the expression of inflammatory cytokine genes including IL-1, IL-6, and TNFα.[4]

c) A3 Calcium channel blockers (CCBs) should help counter the toxic effects of extracellular calcium influx. Although many CCBs are used clinically as anti-arrhythmic and antihypertensive agents, none is currently indicated for treatment of acute CVA. Nimodipine is used to relieve vasospasm secondary to subarachnoid hemorrhage; however, it cannot be given parenterally. Of all the peripheral dihydropyridine CCBs, only nifedipine, nicardipine and A3 can be injected safely by IV.5

d) A4 (GABA-mimetic) – Anecdotal evidence suggests that A2 can induce seizures, especially in the setting of CNS hypoxia.6 Adding an effective anticonvulsant would therefore be prudent. Furthermore, the GABA agonist effects of A4 may be beneficial in blunting glutamate excitotoxicity.

e) AMPA receptor antagonists – Drugs that block this class of glutamate gated ion channels on the postsynaptic membrane have never been used in human trials. One of these agents, a polyamine known as JSTX-3 (a component of Joro spider venom from certain orb weaving spiders native to Japan and Korea) dramatically reduces calcium influx through the AMPA receptor.7 The lack of an effective AMPA receptor blocker may account for the difficulty in minimizing glutamate excitotoxicity using NMDA receptor antagonists or CCBs alone.8

To my knowledge, Joro spider toxin is not lethal in humans. Its mechanism of action is completely different than that of black widow toxin or scorpion toxin. Also, an extensive search of Google and Pubmed revealed zero documented cases of a human death directly attributable to a Joro spider bite. Left to my own devices, I would add a low dose of JSTX (in the micromolar range initially), but keep the crash cart and ventilator handy. In the words of Friedrich Nietzsche (and my med school friend Sean Knapik), "That which does not kill us, only makes us stronger." ;-P

My career in academia and medicine has been atypical, to say the least, but I do not view this negatively. On the contrary, I feel privileged to have learned far more than most people in a multitude of fields including medicine, neurobiology, physiology, and pharmacology. A new approach to treating stroke is an idea I have explored since my undergraduate days at Johns Hopkins University. If I've learned anything from medical school, graduate school, or pharmacy school, it's that I will never abandon my quest to share my interest and insights in the realm of neuroscience with the world.

1. Easton J., Hauser S., and Martin J. (1998) Harrison's Principles of Internal Medicine 14th edition, Cerebrovascular Diseases, Chapter 366: 2325-2348.

2. Sharkey J. and Butcher S.P. (1994) Immunophilins mediate the neuroprotective effects of FK506 in focal cerebral ischemia, Nature, 371: 336-339.

3. Szydlowska K., Zawadzka M., and Kaminska B. (2006) Neuroprotectant FK506 inhibits glutamate-induced apoptosis of astrocytes in vitro and in vivo, Journal of Neurochemistry 99, 965-975.

4. Zawadska M. and Kaminska B. (2005) A Novel mechanism of FK506-mediated neuroprotection: Downregulation of cytokine expression in glial cells, Glia 49 (1), 36-51.

5. Pollack, C. et al (2009) Clevidipine, an IV Dihydropyridine CCB, Is Safe and Effective for the Treatment of Patients with Acute Severe HTN, Annals of Emergency Medicine, Vol. 53(3): 329- 338.

6. Conversation in 2008 with Dr. Rotem Elgavish, Professor of Neurology, U. Alabama at Birmingham.

7. Lino M., Koike M., Isa T., and Ozawa S. (1996) Voltage-dependent blockage of Ca(2+)-permeable AMPA receptors by Joro spider toxin in cultured rat hippocampal neurones. Journal of Physiology, 496 (Part 2): 431–437.

8. Calabresi P. et al (2003) Ionotropic glutamate receptors: still a target for neuroprotection in brain ischemia? Insights from in vitro studies, Neurobiology of Disease, 12: 82-88.

Bryan J. Katz

Religion

Essay 105: Judaism - What is Apocryphal Biblical Literature?

The canonization of the Jewish Bible called the Tanakh in Hebrew and the Old Testament by Christian scholars, began at the end of the Second Temple period around the year 70 CE. Common Era, abbreviated CE, refers to the same time period as A.D. This process probably took well over a century but was essentially complete by the time the Mishnah (Jewish Oral Law) was redacted around the year 218 CE. These two dates are important because they bracket the span of time in which Rabbinic authority was concentrated in the land of Israel.

Historical Context

After the Romans destroyed the Second Temple in the year 70 CE, Rabbis, collectively known as Tanna'im, took it upon themselves to preserve what remained of Jewish life. Part of this effort involved deciding which books of Jewish scripture were sacred and worthy of inclusion in the Tanakh and which writings should be relegated to secondary status. The body of Jewish literature not included by the Rabbis is sometimes referred to as the Apocrypha.

This historical era ended around the year 218 CE when a scholar known as Rav journeyed to Babylonia (modern-day Iraq). With his departure, the center of Jewish scholarship moved to the Babylonian academies, whose opus magnum would be the Talmud Bavli - a 5,500-page compendium of Jewish law, history, fables, and philosophical discussions. Babylonia would become the hub of Jewish life as well as the seat of Rabbinic authority for world Jewry for the next 700 years.

Which major texts were left out of the Jewish Bible and why?

As a general rule, the Rabbis admitted all ancient Jewish texts, including the Book of Daniel, into the third section of the Tanakh, known in Hebrew as the Ketuvim (Writings), or Hagiographia in Greek. Most Bible scholars consider Daniel to be the youngest book of the Old Testament, owing to the presence of several chapters written in Aramaic rather than Biblical Hebrew. The Book of Daniel was ascribed to a body of scholars called the Men of the Great Assembly, who may have compiled it as late as 333 BCE, during the twilight of Persian rule.

That same year, Alexander the Great conquered Judea, the name the Greeks and Romans would later apply to the land of Israel.

Over the next several centuries, Jewish literature was transformed by Judaism's confrontation with the Hellenistic world. The most well-known Jewish works from this period include Baruch (an addendum to the Book of Jeremiah); the Wisdom of Solomon; the Wisdom of Ben Sirach (Ecclesiasticus); First and Second Maccabees; Enoch, Jubilees, Judith, Susanna, and Tobit.

The consensus is that these books were originally written in Hebrew or Aramaic and are almost certainly of Jewish authorship. By this definition, they comprise 'the Apocrypha' in the narrowest sense of the term.

Some consider a collection of miscellaneous works termed the Pseudepigrapha (falsely ascribed writings) to be part of the Apocrypha as well. These books survive as Coptic, Ethiopian, Syriac, or Greek translations; a few may be of Jewish authorship, but most probably are not.

They include the Life of Adam and Eve (also called the Apocalypse or Assumption of Moses), the Testament of the Twelve Patriarchs, and the Testaments of Abraham, Isaac, and Jacob.

Other passages such as Esdras, the Prayer of Manasseh, the Song of the Three Children, and the Destruction of Bel and the Dragon were later additions to older Jewish sources.

Many works of Jewish literature were no doubt lost forever with the destruction of the Second Temple. Others remained hidden for centuries and were only unearthed in modern times. The writings of the Essenes, a Jewish separatist movement whose existence spanned nearly 300 years, were probably unfamiliar to the Rabbis of the first century CE. However, many Essene works were preserved among the Dead Sea Scrolls, which were rediscovered at Qumran (now part of Israel) in 1947.

Generations of scholars have debated the reasons why some books were included in the Tanakh while others were excluded. According to Orthodox Jewish scholars, the explanation is that the Rabbis considered literature written prior to the Hellenistic era to be a divine revelation, or at least divinely inspired, whereas later writings were regarded as neither.

Canonization of the Christian Bible

Christianity was a highly volatile religion for many centuries after its founding. Arguments raged over innumerable aspects of theology, doxology, and, not surprisingly, the contents of the Christian Bible, in particular, the New Testament. Although the details are still a matter of debate, it seems by the time of the Synod of Carthage in 397 CE, all of the books now considered part of the New Testament had been given canonical status by the Roman Catholic Church.

By the early 5th century, most Eastern Orthodox churches followed suit. The four gospels of Matthew, Mark, Luke, and John became the core of the Christian Bible. Several other gospels, such as Peter, Thomas, Judas, and Mary Magdalene, were excluded. These and other writings of an early Christian sect called the Gnostics were discovered at Nag Hammadi, Egypt, in 1945.

Interestingly, the Roman Catholic and Eastern Orthodox churches conferred canonical status on certain Apocryphal books, including the Wisdom of Solomon, the Wisdom of Ben Sirach, Judith, Maccabees, as well as the aforementioned additions to the Books of Esther, Daniel, and Chronicles. The Ethiopian Orthodox Church includes Enoch, Jubilees, and a three-volume work called the Book of Meqabyan in its canon. Protestant churches largely reject these Apocryphal works as inappropriate additions to the Bible.

Essay 106: Judaism - The Patriarchs

The story of the Three Patriarchs of Judaism (Abraham, Isaac, and Jacob) is found in the Biblical Book of Genesis, known in Hebrew as Sefer Bereshit. Although archaeologists have yet to find evidence of these three men outside of the Bible, one should remember that few records from this ancient period of Middle Eastern history survive aside from the pharaohs' tombs in Egypt and the Code of Hammurabi in Mesopotamia. At any rate, the only way to understand the importance of the Patriarchs to the development of Judaism is to examine the life of each one.

Abraham is considered the first Jew and, by extension, the founder of monotheism itself. Centuries after he lived, Jewish scholars wrote down stories endeavoring to explain how Abraham, the son of a Babylonian idol carver named Terach, discovered God on his own when he was a small child.

The most popular account claims that Abraham observed how the sun and the moon ruled the day and night, respectively, but neither celestial body was all-powerful. He then deduced that a higher power (God) must be in charge behind the scenes. After this epiphany, Abraham smashed his father's idols except for the largest. He placed the club in the hands of the largest idol, then told his father that this idol had grown angry and proceeded to smash all of the smaller idols to bits. When his father protested that idols were inanimate objects, Abraham replied by asking why his father carved and worshiped them. Some versions of the story claim that Terach immediately abandoned idolatry; others leave the outcome of the story to the reader's imagination.

Volumes of commentaries exist that scrutinize every detail of Abraham's life as presented in the Bible; further volumes of apocryphal literature fill in every imaginable detail. Although a Biblical analysis is beyond the scope of this article, the highlights of Abraham's life include his migration to the land of the Canaanites (destined to become the Land of Israel), his journeys to Egypt and the Philistine kingdoms to escape famine, the rift with his nephew Lot; his victory in the War of the Four Kings against the Five; his meeting with the three angels; fathering Ishmael by Hagar and Isaac by Sarah; the test of expelling Ishmael and Hagar from his home; his final test of binding Isaac as a sacrifice on Mount Moriah (called the Akedat Yitzchak); his purchase of the Cave of Machpelah for the burial of his wife Sarah; fathering several more children near the end of his life by a woman called Keturah (identified in Rabbinic literature with Hagar); and finally, his death at the supernatural age of 175.

The Talmud in tractate Berachot attributes the Jewish morning prayer service to Abraham when he awoke at dawn to pray for his nephew Lot's safety after Sodom's destruction. Medieval rabbis ascribed Psalm 89 to Abraham (written under the pseudonym Eitan Haezrachi). In later Kabbalistic literature, Abraham is described as the epitome of kindness, or Chesed, as it is known in Hebrew.

The Book of Genesis devotes far less time to the life of Isaac other than a few events outside of those connected to his father, Abraham. Isaac married his cousin Rebecca three years after the Akedah at age 40.

Twenty years later, his sons Jacob and Esau were born. Beyond that, the Bible relates that Isaac excavated the wells dug by Abraham, his journey to the land of the Philistines, his blindness in old age, his awarding the blessings of the firstborn to Jacob rather than Esau following the deception by Rebecca and Jacob; and his death at the age of 180. The Talmud ascribes the Jewish afternoon prayer service to Isaac. Kabbalistic literature identifies Isaac with the attribute of restraint, or Gevurah, epitomized by the Akedah described in Genesis 22:1-19.

Of the three patriarchs, the Bible spends the most time discussing the life of Jacob. Jacob's name was notably changed to Israel after he wrestled with an angel, as described in Genesis 32:25-33. Other key events in Jacob's life were his escape from his murderous brother Esau, his marriage to his cousins Rachel and Leah, fathering twelve sons and one daughter by his two wives and their two handmaids, his escape from his father-in-law Laban; his return to Canaan as a set up for the death of Rachel; the sale of Joseph by his brothers; the migration of Jacob and his family to Egypt during a famine; their reunion with Joseph, now viceroy of Egypt; and Jacob's death at the age of 147.

Because Jacob's life was marked by far more trials and tribulations than those of his father or grandfather, Rabbinic commentators transformed him into a microcosm of the Jewish people's struggle for survival against overwhelming odds. The Talmud ascribes the Jewish evening prayer service to Jacob, specifically to his vision of the ladder in Genesis 27:11-14. In Kabbalistic literature, Jacob is identified with the attribute of splendor, or Tiferet, which strikes a balance between the opposite qualities of kindness and restraint represented by Abraham and Isaac, respectively.

Essay 107: Judaism - The Star of David

For centuries, the hexagram, or six-pointed star, has been used as a symbol of the Jewish people and, by extension, Judaism. Although the popular name for this symbol is the Star of David, it is known as the Mogen David or Shield of David in Hebrew. According to some historians, Jewish warriors led by King David painted their shields with this symbol over 3,000 years ago. Since the Bible itself (in particular, the Books of Samuel and Kings) makes no direct mention of the Star of David, there is little consensus as to the origin of this symbol. Naturally, several alternative theories have been suggested.

The late Jewish historian Chaim Potok asserted that merchants commonly used the Star of David throughout the Near East during ancient times; however, its origin remains obscure, and its original meaning has long since been lost to history. Other theologians suggest that the six points of the star symbolize God's control over the six cardinal directions of the physical world.

Several years ago, a tour guide told me that the Star of David originated in India and that the two opposing triangles symbolize humanity's free will to choose good or evil. This idea is similar to the Kabbalistic (Jewish mystical) explanation given in this and the previous reference. As for the Indian provenance of the Star of David, scant evidence exists to support or refute this claim.

One of the most interesting theories devised to explain the meaning of the Star of David was formulated by Franz Rosenzweig in his book "The Star of Redemption," written during World War I. Briefly, the three vertices of the right side up triangle represent God, humanity, and the universe. According to Rosenzweig, the importance of all three entities lies not in their abstract meanings but rather in their relationships with each other.

As such, it follows that the three points of the upside-down triangle represent the relationships between each member of this triad. God's relationship to the universe is termed Creation; His relationship to humanity is expressed as Revelation; and humanity's relationship to the universe (or the purpose of existence) is Redemption, although a more accurate translation would be the perfection or elevation of the physical world back to its spiritual source.

In 1948, the Star of David was officially chosen as the symbol of the State of Israel when it declared its independence from British rule. The two blue stripes on the Israeli flag represent the miraculous splitting of the Red Sea described in the Book of Exodus. The Star of David symbolizes the safe passage of the Jewish people through the parted Red Sea as they escaped from slavery in Egypt.

Travel

Essay 108: A Whirlwind Tour of Destinations

The Caribbean and Mexico offer an amazing variety of enjoyable as well as educational places to visit. An important consideration for prospective visitors, especially those from the U.S. and Canada, is to keep in mind that, aside from travel costs, most of these places are affordable and (aside from parts of Mexico) comparatively safe - a rare combination nowadays.

The eastern Caribbean consists of the Bahamas, Puerto Rico, and the islands of the Lesser Antilles. The Bahamas gained its independence from Great Britain in 1973. In the four decades since then, the entire economy of these islands has revolved around tourism, really an umbrella term for everything from hotel-casinos to luxury beach resorts to snorkeling and deep-sea fishing. The most famous resort in the Bahamas is probably the Atlantis. This magnificent hotel was completed in 1998 and features a huge aquarium directly connected to the ocean.

Puerto Rico is the only U.S. commonwealth in the Caribbean. This island was captured from Spain in 1898 during the Spanish-American War and has thus far chosen to remain a commonwealth as opposed to seeking full independence or becoming America's 51st state. The capital of Puerto Rico, San Juan, has some interesting historical sites, particularly a large Spanish fort administered by the U.S. Park Service. In addition to beautiful beaches, Puerto Rico has a protected rainforest preserve. There are also a number of coffee plantations that export coffee beans directly to the Vatican.

The best-known islands of the Lesser Antilles include Antigua, Barbuda, Guadeloupe, Martinique, Dominica, St. Lucia, Grenada, Barbados, St. Maarten, Trinidad and Tobago, and the Virgin Islands. Antigua and Barbuda became independent from England in the 1980s. Aside from a few gated resorts, much of Antigua was covered by dilapidated shacks and herds of goats, at least as of 1988; in all likelihood, not much has changed since then.

Guadeloupe and Martinique are overseas departments of France. Martinique has an active volcano that erupted in 1902 and destroyed much of the island. One of the few survivors was in prison at the time and became a local hero after being found alive several days later. Some of the former plantations have become historical sites with restaurants and shops to attract tourists.

Dominica became independent of British rule in 1988. Much of the island is still covered by lush rainforest carpeting its steep slopes and fern gullies. One of the Pirates of the Caribbean movies was filmed here a few years ago.

Barbados was controlled at different times by the Spanish and British, achieving independence in 1966. As with most Caribbean islands, the population is largely Catholic, and the entire island shuts down from Good Friday to Easter Sunday. Barbados also claims to house the oldest synagogue in the Western Hemisphere, built in 1654. St. Lucia is one of the closest islands to Barbados and remains a British territory.

St. Maarten is divided between French and Dutch control. Legend has it that rather than fighting over this small island, the French and Dutch instructed surveyors to walk in opposite directions along the coast until they met on the opposite side of the island. In this way, the French ended up with slightly more of St. Maarten than the Dutch.

Three of the Virgin Islands are U.S. territory; the others are British. The U.S. purchased these islands from Denmark near the end of WWI; the two most populous are St. Thomas and St. Croix. St. Thomas was once the lair of many famous pirates, including Bluebeard, Blackbeard, and several others. This island also boasts a famous underwater aquarium, well-kept beaches, and many picturesque homes perched atop the hills.

Even though this island is a U.S. territory, motorists still drive on the left side of the road. Be aware of this if you decide to rent a car in St. Thomas.

Trinidad and Tobago are located off the coast of Venezuela. These islands were under Spanish rule until 1797, then British control, finally gaining their independence in 1962. Laying to the west of the Lesser Antilles, the islands of the southern Caribbean consist of the Grand Caymans, a few other British territories, and the Netherlands Antilles, which includes the nominally independent islands of Aruba, Bonaire, and Curacao. These three are small Leeward Islands that receive relatively little rainfall. Their economies are based on tourism, oil refineries, and rum distilleries.

The Western Caribbean (or islands not already covered) consists of the Greater Antilles - Cuba, Hispaniola, and Jamaica; as well as a few islands belonging to Mexico, particularly Cozumel. Direct travel to Cuba is not an option for American tourists; however, flights are available from Mexico, the Dominican Republic, and most other Latin American countries. The island of Hispaniola consists of Haiti and the Dominican Republic. Tourism to Haiti, already in decline for many years, has come to a standstill after the devastating earthquake of 2010. In contrast, tourism to the Dominican Republic is on the rise. Finally, Jamaica remains a popular tourist destination, particularly Dunn's River Falls in Ocho Rios and the beach resorts in Montego Bay. This island also boasts one of the world's rarest, most expensive coffees - Jamaica Blue Mountain coffee.

Compared to the Caribbean, Mexico is far larger in terms of its geographic size and population. As far as tourist destinations go, however, if your top priority is safety, you're no doubt aware that most of Mexico is far less safe today than it was even 5 years ago. Tourists are advised to avoid the northern half of Mexico entirely in light of the ongoing battles between the Mexican government and drug cartels in Juarez, Tijuana, and numerous other places.

The capital of Mexico is Mexico City, located over 1,000 miles south of El Paso, Texas. It is one of the largest cities in the world today, home to well over 15 million people. Unfortunately, Mexico City also suffers from some of the worst air pollution in the world, considering its cars and taxi cabs number in the millions. Mexico City was built on the site of the ancient Aztec capital of Tenochtitlan, conquered by the Spaniards under Cortez in 1521. Famous landmarks include El Zocalo or central plaza, the site of the Temple Mayor, the National Palace, several museums, and cathedrals dating to the 16th century.

Beyond Mexico's sprawling capital, other highlights include the so-called Mexican Riviera (mainly the cities of Acapulco, Puerto Vallarta, and Cabo San Lucas) and the Yucatan peninsula, arguably the safest part of the country for tourists. The Yucatan's most popular attractions include the resorts of Cancun, a nature park called Xcaret, and the Mayan ruins of Chichen Itza.

One of Mexico's finest national parks, Chichen Itza, is possibly the best preserved Mayan site outside Guatemala. At the heart of Chichen Itza stands a large pyramid called El Castillo in Spanish. The pyramid has 91 steps on each side and an obelisk on top, corresponding to the 365 days of the solar calendar. Twice a year, on the spring and fall equinoxes, thousands gather to watch as the sun's rays illuminate the feathered serpent carvings on the pyramid. Excavations at Chichen Itza have turned up a ball court, an astronomical observatory, and many other structures whose original purpose is still a subject of debate among archaeologists.

Essay 109: Fall Foliage - MD, DC, VA

When it comes to destinations for seeing beautiful fall foliage in the Mid-Atlantic states, you have a tremendous number of options to choose from. Here are some of the most popular attractions, along with a few of the region's less familiar treasures.

Maryland

Patapsco State Park

Far and away, this is the largest forested area remaining along the I-95 corridor connecting Baltimore and Washington, DC. The park features hiking and mountain biking trails, fishing, canoeing, and horseback riding.

Western Maryland: The Appalachian Trail and Deep Creek Lake State Park

Most of Maryland's intact forests are located between Hagerstown and West Virginia. The Appalachian Trail traverses 41 miles of western Maryland and is extremely popular with hikers. Deep Creek Lake is actually a manmade lake completed as part of a hydroelectric dam project in 1925. Although most visitors come during the summer months for swimming and boating, there is no doubt the fall foliage is spectacular.

Gunpowder Falls State Park

This state park is one of the best-preserved forests between Baltimore and Delaware. Part of the Chesapeake Bay watershed, the park is home to an abundance of birds, fish, and wildlife.

Washington D.C.

Best known for its Cherry Blossom Festival during the spring, your best bet for seeing autumn beauty in our nation's capital will probably be around the Tidal Basin, a few local parks, the National Zoo, and university campuses like Georgetown and American University. Most of the District is a relatively small, densely populated urban area, as opposed to a pristine forest. The main outdoor activities here include jogging, hiking, biking around the Tidal Basin, and trips to the aforementioned areas frequented by tourists.

Virginia

One of the most popular wilderness attractions in Virginia is Skyline Drive. This corridor runs over 100 miles from northeast to southwest and is within easy driving distance from I-81. Skyline Drive is the gateway to Virginia's most scenic area, the Shenandoah Valley.

Shenandoah National Park

This lush, forested area marks the southern end of Skyline Drive. Immensely popular with tourists, it was designated a National Park in 1935. The Shenandoah Valley is also well known for its wineries, bed and breakfasts, and Civil War battle sites.

George Washington National Forest

Tucked away near the border with West Virginia, this forest forms part of Virginia's largest, most pristine wilderness area west of the Blue Ridge Mountains, rivaling the Great Smoky Mountains in its magnificent beauty.

Essay 110: High School Memories: My Senior Class Trip To Israel

One of the main reasons people stuck around for high school at Beth Tfiloh was the trip to Israel promised to every senior class. Actually, the class of 1991 did not go to Israel because Gulf War I was about to happen. Ironically, if they had gone in January as they had planned, they would have been back home a day before the war started. Oh well.

Our class spent eighteen days in Israel in March 1992, and I have to say they were eighteen of the happiest days of my life. Partially because I was away from home with my friends but mainly because for someone who has spent his entire life learning all about Israel, going there for the first time is truly an awesome experience.

That was over 30 years ago, but sometimes it feels like yesterday. And truth be told, I wish the situation today in Israel were as rosy as it was back in 1992. For those readers who haven't been to Israel, I'll set the scene. We flew out of JFK airport in New York at midnight and reached Ben Gurion Airport that afternoon. Everyone was tired and sore from the 10 hour plane ride, but I was excited. We were going directly from the airport to pray at the Kotel, the Wailing Wall in the Old City of Jerusalem. As physically exhausted as I felt, I couldn't wait to see Jerusalem in real life.

The bus finally pulled up to the walls of the Old City, and we walked through the Zion Gate then down a long staircase to the Wailing Wall. The moment I caught sight of the Temple Mount, I felt this vibe – a mixture of joy, tranquility, goodness, and power combined – like G-d's Presence was right there welcoming everyone. And as I touched the stones of the Wall, the feeling that life is precious and wonderful and flows from G-d Himself cannot be expressed but only experienced. Seriously – everyone out there regardless of religion, or lack thereof, can experience this feeling. You have to be there in real life to know what I mean, but trust me, it is so worth it.

And for the next three weeks, we experienced this every single day. One concrete example of this was the way we treated the people we met and how everyone treated one another. Whenever panhandlers approached us for change, someone always managed to come up with a few coins. No one thought to ask them, "Are you Jewish? Are you Arab?" It didn't matter. And all the people we encountered – a diverse, opinionated bunch – seemed genuinely friendly. Cynics would say the reason was that we were American tourists, and the locals saw dollar signs. But truthfully, the money we spent in three weeks would have lasted all of one weekend in New York or L.A.

Literally everyone we ran into seemed genuinely happy. Religious Jews, secular Jews, and, yes, the Arabs in Israel and Judea-Samaria all seemed positive and upbeat. Keep in mind that this was less than five years after the 1987 Intifada – which the world media portrayed as a permanent wedge dividing Jews and Arabs into hopelessly irreconcilable camps.

The media is usually lousy at predicting the future, but many years later, their self-fulfilling prophecy came true. Which I'm sure made them gleeful – because deep down, most media people are bloodthirsty vultures who enjoy broadcasting war stories from the safety of a news studio 6,000 miles away. But I'm getting ahead of myself. I'll come back to the current mess at the end of the chapter. For now, let's get back to the story.

We spent the first week of the trip visiting places in and around Jerusalem. My favorite trip, after the Western Wall, of course, was this museum dedicated to the history of Jerusalem.

Viewing the millennia of history connecting us to the Jews of King David's time who made Jerusalem the center of Jewish existence made me appreciate how history has come full circle. After all is said and done, the Jews survived 2,000 years of exile and persecution to reestablish the state of Israel in 1948 and go on to liberate all of Jerusalem in 1967.

That's no small feat. And for anyone who focuses on the big picture, I think you'll conclude that the story of the Jewish people collectively and the history of Jerusalem as a microcosm of the Jewish experience is an eternal source of inspiration for all mankind. Here's a people and a city that were overrun dozens of times in the past 3,000 years – by Babylonians, Greeks, Syrians, Romans, Byzantines, Arabs, Crusaders, Mongols, and Ottoman Turks. Paradoxically, all of these groups have declined or vanished into history with the exception of the Jews themselves. And the only explanation that has stood the test of time is G-d's promise that both the Jews and Jerusalem are a flame that can never be extinguished.

From Jerusalem, we went to the Galilee in northern Israel. The two main destinations there are Safed and Tiberias, near the Sea of Galilee (actually a lake). Tiberias is famous for containing the graves of many ancient and medieval Rabbinic sages. In recent decades it has become an artists' colony. Safed (Tzefat in Hebrew) had some war memorial and not much else that I remember. We also went to the Israeli-Lebanese border, which at that time was the civilian border because Israel maintained a ten mile security zone inside southern Lebanon.

The last phase of the trip took us to the Negev, where the cities of Be'ersheva and Arad are located. We also passed by S'de Boker, where Israel's first prime minister, David Ben Gurion, spent the last years of his life. We went on a hike in some desert canyon called Ein Avdot. Instead of following the path, a few class members, led by Allon, started climbing directly down the rocky slope. This prompted our tour guide Yaakov to scream, "Get off the bloody cliff!" I know it doesn't sound very funny. It's one of those times you had to be there to know how funny it really was.

Those were the good old days. Fast forward 32 years, and life is drastically different in Israel. The years since the ill-fated Oslo Accords of 1993 have driven home two fundamental lessons that Israelis must never forget:

1) It takes two sides to make peace but only one to make war.

2) Negotiating with unrepentant terrorists (Arafat/Hamas), is a slap in the face to all innocent victims of Palestinian terrorism.

Let me end this chapter with a story that puts the conflict in perspective. There's a parable in the Talmud about a group of people traveling on board a ship. All of a sudden, someone gets the bright idea to start drilling a hole underneath his seat. As the water begins rushing in, the rest of the passengers start shouting, "Are you crazy? What are you doing? You'll drown us all!"

To which the man responds, "Mind your own business. I'm only drilling under my seat."

And that sums up the situation quite well – specifically for the Arabs held hostage by their own terrorist fighters and suicide bombers. They are drilling a massive hole, the ship is flooding, and a lot of innocent people are going to drown. One glance at Gaza tells you all you need to know…

U.S. History & the American Experience

Essay 111: Economics 101: Dr. Katz Goes to Wall Street

I am not here to rewrite an economics textbook. My point is to move beyond concepts like supply and demand (the obsession of microeconomics) and give you, the reader, insight into the big picture. Because I'm a big-picture kind of guy!! Also, I took macroeconomics at Hopkins but never got around to taking microeconomics. Honestly, I sat through one micro lecture and nearly fell asleep. The professor drew a couple of graphs that purported to show why it was a good idea to grow corn in Iowa and wheat in Montana as opposed to the other way around. Since that sort of information is both glaringly obvious and useless to the public at large, here's my somewhat satirical take on Economics 101.

Point I: The Sea of Debt – Relax and Float. Some people are shocked to hear that the global economy floats on a virtually bottomless sea of debt. These are the same people who thought the world was coming to an end when the year 2000 was upon us, and they'll probably be hiding in their basements until the end of 2012, even though the Mayans in Guatemala don't seem the least bit worried. If I have taught you, the reader, anything so far, it's that the only tangible results of a crisis mentality are peptic ulcers. And A-bomb fallout shelters, but I digress.

The sea of debt is a simple fact of life. Most individuals who continue their education beyond high school incur debt. Corporations go into debt as part of their long-term growth strategy. Most countries in the world owe a debt to other countries. The U.S. government is close to $17 trillion in debt. And the sky hasn't fallen. Here's the take-home point:

If everyone owes money to everyone else, at the end of the day, it all cancels out!

Why? If everyone called in his/her debts at the same time – essentially pulling the plug – the sea of debt would empty, and everyone would go down the drain. It does not require a leap of logic to understand that no one wants this to happen.

Think back to the early 1980s when the Pac-Man era was in full swing. When you lost your last Pac-Man to the ghost monsters, the machine made a sound like wawa-wawa-woink-woink, and two dreaded words appeared: Game Over.

Imagine how pissed off you were at having to spend another quarter to start the game all over again. Now imagine that instead of losing twenty-five cents, the game had cost you your entire life savings. Multiply that feeling by 7 billion, and you'll get an inkling of how the human race would feel in the wake of a global economic collapse. Pretty damn pissed!

That's why, in the years since the Great Depression, we've wised up. For starters, bank accounts are insured up to $100,000 (assuming the bank is a member of FDIC).

Second, we've abandoned antiquated concepts like the gold standard and converted tangible money into intangible numbers on computers. At this point, money really does become as worthless as the paper it's printed on. The only thing propping up the global financial system is, to borrow a phrase from Rousseau, the general will that the show must go on.

Point II: Job Security. I've got good news and bad news when it comes to the job market. The good news is that at least three employment fields offer absolute, lifetime job security.

The bad news is that your choices are limited to the U.S. Postal Service, the Department of Motor Vehicles, or Janitorial Services, a.k.a. Sanitation Engineer if you're into euphemisms.

While I'm on the subject, will someone explain to me why postal workers "go postal"? From what I can tell, working at the post office is one of the least stressful jobs I can imagine. The only job requirement seems to be showing up for work wearing clothing of some kind. You can be as slow and inefficient as you want; after all, many post offices come equipped with bulletproof glass in front of every booth. No matter how long the line gets, only half the booths are ever staffed. About all the customers can do is shake their heads and mutter curse words.

The DMV is even better. You can show up, eat doughnuts, drink coffee, smoke cigarettes, accomplish absolutely nothing, and still not get fired. If you feel like doing something, you can glare at the people in line, blink occasionally, or mumble vague directions regarding the whereabouts of important forms and applications. If someone asks you where the restroom is located, say, "at the end of the hall," then abruptly walk away. Remember, as a state employee, it takes about a pound of paperwork to fire someone, and no one feels like getting carpal tunnel syndrome in the process.

As a janitor/sanitation engineer, you'll actually have to do real work, but there are certain perks: a lifetime supply of free cardboard boxes and cleaning supplies, free rides on Zamboni machines, and the G-d given right to yell at anyone who dares walk across the floor when it is wet, including the company CEO.

Point III: When it comes to the future of Social Security and Medicare, don't sweat it. Since the federal government treats money like it's playing a game of monopoly, it will shovel money into these two programs forever – meaning until people stop growing old or until the day Earth crashes into the sun.

Point IV: Taxes. I'm sure you've heard Benjamin Franklin's saying that nothing in this world is certain except for death and taxes. That may be true. Same with the expression about the IRS:

They've got what it takes to take what you've got. But instead of thinking of taxes as a punishment, think of them as a challenge. How can you conceal or transfer enough of your assets to avoid paying any substantial income tax? Here's some food for thought:

(A) Employ children and friends as proxies to purchase appreciating items like real estate, precious metals, and gems. You can claim your children as dependents, and your friends will think it's exciting – especially if they get a cut.

(B) If your annual income exceeds $100,000, begin transferring money into overseas accounts in Switzerland or, better yet, the Grand Cayman Islands. Did you know that the Cayman Islands are the fourth largest banking center in the world? Would you like to guess why? First hint: they're not far from Colombia. Second hint: They have a Clintonian policy of don't ask, don't tell. Remember, if the Medellín cartel can stash its money there, so can you.

(C) Avoid most stocks except for blue-chip stocks like Microsoft, Walmart, and Coca-Cola. Put those in your children's names. Avoid stock options like the plague unless you can dump them at a moment's notice. A good friend of mine learned this lesson the hard way and ended up owing the IRS over 1 million dollars.

(D) If you are religious, consider filing for tax-exempt status as a self-proclaimed synagogue/church/temple/cult leader. This will work if and only if your friends/followers corroborate your story. The downside is that total tax evasion will still land you in jail; remember what happened to Reverend Sun Myung Moon in 1983? The upshot is that your followers may pledge their eternal loyalty and give you a donation whenever you tell them God needs money.

(E) If all else fails, and you find yourself buried under a mountain of debt, consider assuming a new identity. Bonus points if you can convincingly fake your own death. Most credit card companies will cancel debts up to $25,000 when presented with a valid death certificate. For pointers, watch the 1986 cult classic FX.

(F) When it comes to paying taxes, you may opt to use a service like H&R Block. This saves you the time and headache of filling out a tax form. Tax forms are so convoluted and confusing it's like reading the manual of a Simon Says game for accountants. Add lines A and B, subtract line C, multiply by the square root of line D, and then skip to line G if you own an orange grove in Florida.

But buyer beware. According to my father, the people at H&R Block are average Joes with 2-3 weeks of training under their belts. My father also claims the IRS will send a representative out to help you do your tax returns, but in my opinion, that's inviting an audit.

(G) Here's the Dr. Katz approach to taxes: Either pester a knowledgeable relative with enough patience to do your taxes or, if you're so inclined, try it yourself. First, get a 1040 EZ form from the post office. Next, fill in the after-tax amount listed on your W-2 form, then deduct tax credits for any dependents as well as donations to charity, and finally, sign and mail. If you own stocks, add the dividends earned to your income.

Since many stocks have plummeted in value in recent years, this should not make much difference. Be creative with deductions. "Charity" has a very broad definition - anytime you buy anything for anyone who can't afford it – including homeless people. Just remember, if your annual income is below $100,000, don't sweat it; the IRS has bigger fish to fry.

Point V: Most life insurance policies are scams. First off, let's call a spade a spade. It's death insurance. After all, the company pays nothing as long as the policyholder is alive (or lives to age 100). Life insurance is the ultimate example of the Law of the Font, i.e., the big print giveth; the small print taketh away.

Remember this the major life insurance companies are largely run by crooks who try to withhold every penny they can from their rightful owners.

How else can these people afford mansions, expensive cars, and monthly trips to Bora Bora? The take-home point: if you feel compelled to buy a life insurance policy, do NOT sign it until a skilled attorney goes over it with a fine-toothed comb.

I admit that some forms of insurance are legally necessary, like auto insurance, and others are practically essential, like health insurance. The key to buying insurance is knowing how to derive the maximum benefit from your policy. After all, you're the one paying for it!

Let's look at the auto insurance racket. This industry is run by a bunch of thieves who will jack up your insurance premiums anytime you report anything to them. Which really sucks ass. If you're in an accident, and it was the other guy's fault, they'll still raise your rates.

Why? Because they're a bunch of greedy bastards. No more or less profound than that. Considering how much they charge for auto insurance in the first place, never hesitate to attribute any damage done to your vehicle to unwitnessed acts of vandalism or hit-and-run accidents. There's really no way a claims adjuster can disprove your story. You can have several thousand dollars of bodywork done to your car for the bargain price of your deductible, around $250. Some people call this insurance fraud; I call it thinking.

In the case of health insurance, it's a classic case of not what you know but who you know.

If you happen to know a physician, do not hesitate to ask his/her advice about any medical questions or health concerns you may have. They should remember something from medical school. If you're hoping for free samples of medications, forget about asking a physician.

Most consider it to be stealing even if they sign out a sample under their own names, which I honestly don't understand, considering that the clinic or hospital either got the medicine for free from a pharmaceutical representative or else they paid cents on the dollar for it.

Unless you have connections at a subsidized clinic, a pharmacist is your point of contact for discounted over-the-counter medicine. They order vast quantities of medication every week. And with the understandable exception of controlled substances like narcotics and sleeping pills, the FDA and DEA couldn't care less who gets their hands on what.

If this country were as progressive as Mexico, every oral medication except opioids, amphetamines, and sedatives/hypnotics would be sold over the counter. That would save many unnecessary trips to the doctor for prescription refills. And it would save doctors lots of annoying paperwork and calls to the pharmacy ☐

Point VI: It's time to put the funeral industry out of business. This is one expense I do not want to incur personally or pass on to any of my descendants. Supposedly, the average funeral today costs over $7,000. Can someone please explain to me how a wooden coffin and a six-foot hole in the ground can cost that much? I suppose you have to figure in other fees like labor, the cost of the cemetery plot, and the tombstone. Still, the question remains: why shell out all that money when the person is dead and can't possibly appreciate it?

Here's my solution: do it yourself for funerals. Instead of a pre-assembled coffin, they should sell the materials at a place like Ikea (Scandinavian for junkyard). Or better yet, explain what happened to the people at Home Depot and get a sturdy cardboard box for free. Some Hefty bags and a roll of duct tape ought to do that. The ceremony can take place in the living room, and the burial can be held in the backyard. For those of you who live in an apartment building, a backyard burial will not be an option. Instead, consider cremation or burial at sea.

If cremation conflicts with your religious beliefs (as it does with mine since I'm Jewish), consider waiting until the middle of the night and then burying the deceased person in the woods or in a remote area of some large park. You can erect a makeshift memorial later, and assuming you dig the grave deep enough, no one will ever know.

Point VII: The Dow Jones Industrial Average is a virtually meaningless measurement of the health of the U.S. economy. I'm not saying that the DJIA is totally inaccurate, only that the day-to-day performance of thirty companies yields an incomplete picture of the world as it really is. Better indicators of the nation's economic health would be purchases of new homes and cars, money spent on travel and recreation, and the number of people enrolled in colleges and universities.

While we're on the subject of gauging the health of the U.S. economy, let's review the fundamental equation of macroeconomics. First, a word of caution. The validity of this equation rests on the assumption that a nation's income is indeed equal to its economic output. I believe this idea is a cornerstone of Keynesian economic theory. I also suspect that most economists younger than Alan Greenspan have never read John Maynard Keynes.

Admittedly, Gross National Product (GNP) is an imperfect measure of a nation's economic output. For example, it ignores the existence of the black market - drug trafficking in Colombia, weapons smuggling in Russia and the Middle East, babies from Romania, and kidneys from India and China. Nevertheless, accurately analyzing the components of GNP captures the strengths and weaknesses of ANY nation's economy, even a garbage can like North Korea, where twigs and worms are the staple foods.

Here's the equation, along with a brief explanation:

GNP = C + I + G + (X-IM)

C denotes consumer spending; I includes private investment in factories and machinery as well as purchases of new homes; G refers to government spending; finally, X-IM is shorthand for the balance of international trade or the overall value of exports minus imports.

All conceivable economic activity (even the illegal stuff) falls into one of these four categories. As with Einstein's Theory of Relativity, no exceptions to this rule have ever been found. I will readily admit that I have no idea who discovered this simple but powerful equation at the core of macroeconomics, but s/he deserves the Nobel Prize more than any economist alive today.

Point VIII: Prices. According to my economics textbook, prices are what they are because someone is willing to pay that much for the item in question. Stated another way, the point at which an item's supply and demand curves intersect determines the price level. Sounds easy enough. I want to remind people of a simple fact. If enough people boycotted expensive things, their prices would fall. It's called using the demand curve in our favor. We should do it more often.

Point IX: The Golden Rule – Myth and Reality. In the realm of economics, so the saying goes, s/he who has the gold rules. In the past, this was absolutely true. However, since the U.S. went off the gold standard, gold has become a commodity like oil, coffee, or cocoa beans. It amazes me that it took a worldwide depression for it to dawn on people that although gold is shiny and makes nice jewelry, it cannot be consumed as food or used to heat one's dwelling or fuel one's automobile. Hence, the "golden rule" became secondary to the general will, i.e., that floating on the sea of debt is individuals' and nations' top priority.

As far as giant corporations go, while they do wield tremendous power in the U.S. economy, they are ultimately at the mercy of the law of supply and demand. In addition, there is always the possibility of hauling a mega-corporation into court for breach of antitrust laws. Thus far, Standard Oil and AT&T have been the only monopolies successfully broken up under the Sherman Antitrust Act. Microsoft almost met this fate; evidently, Bill Gates persuaded President Bush to persuade the Supreme Court to drop the case. Microsoft survived intact, but Gates lost $50 billion in the protracted legal mess, not to mention the ensuing dot com collapse of 2000 - 2001.

Recently, this new operating system called Linux emerged as a competitor with Microsoft, which technically means that Microsoft Windows no longer has a monopoly on the software market. But no one uses Linux, except for one Ph.D. researcher I know at the University of Alabama. Bill Gates is still officially the richest person in the world (my father remains convinced that the late King Fahd of Saudi Arabia was richer), and he serves as the exception that proves the rule.

Point X: Credit, credit cards, and creditors. First, the only things you need to obtain credit in the USA are a pulse and a signature. I'm totally serious. Credit card companies know that over half of their customers never pay off the full balance each month – which guarantees the company will make a huge profit, if only from finance charges. This brings me to my second point: Avoid credit card debt like the plague. Having said that, credit card debt need not be fatal; rather, like a chronic disease, credit card debt is manageable if you employ certain tactics.

1)   Consolidate the debt as much as possible onto one card; this will help avoid high-interest rates and finance charges, but if and only if:

2)   You take advantage of those 0% APR (Annual Percentage Rate) offers in the mail. This translates into six or even twelve months of interest-free repayment time. An added bonus is free balance transfers to the 0% APR card, but these are few and far between. Nonetheless, most cards cap the balance transfer fee at $75 – which isn't a big deal in the grand scheme of things.

3)   Go out of your way to be nice to the customer service reps over the phone. If you are convincing at telling them hard luck stories, they will happily waive late fees and reduce the APR on your balance.

4)   In the unhappy event that the credit card people add you to their naughty list, you must be ready to fight fire with fire. First, get caller ID ASAP.

Next, convince the boobs in customer service that the person they are after has (take your pick) lost their job / lost their mind / become a member of a cult, and ceased all contact with the outside world. Reserve the ultimate creative story (debtor is deceased) for $25,000 or more bills. Wait at least six months before applying for credit cards again. If they insist you are dead, ask them in a flabbergasted tone of voice what the hell they're talking about.

Point XI: Bankruptcy – the last resort. Although people have long dreaded the "B" word, the truth is over half a million people file for bankruptcy every year. The downside is they take away all of your credit cards for five years or more. Also, bankruptcy does not cover student loans or taxes owed to the IRS. The upshot is that it gives you time to get back on your feet without being hounded by semiliterate goons from the credit bureau. They'll eventually get it through their very thick skulls that you can't collect much money from someone with no assets.

Smart people will manage to transfer every asset imaginable to friends or family so that they own nothing on paper if and when they file for bankruptcy. Now that I think about it, most cool people have declared bankruptcy at least once. Like Donald Trump. Or this millionaire financial planner from Baltimore, Julius Westheimer, who supposedly declared bankruptcy three or four times. And WBAL radio still invited him to give on-air financial advice until his death in 2005.

On the international scene, virtually every country in Latin America and sub-Saharan Africa has perpetual bankruptcy. This stems from a nearly endless cycle of civil wars in addition to what passes for governments in these countries periodically giving the middle finger to the World Bank and IMF.

Initially, this sounds counterproductive, but these countries may be on to something in the long run. In essence, they gave up paying off their debts long ago; instead, they set the calculator back to zero whenever the number got too large to fit on the screen. Lucky for them, they have the old model calculators that don't have exponential notation.

The combination of inhospitable environments and lack of modern technology actually works to these countries' advantage. Would anyone in his right mind trek across the Sahara or into the Amazon to politely remind a warlord and his gangster army that their debt payments are way past due? At best, s/he would get laughed at; at worst, they might get eaten alive by mosquitoes, crocodiles, or starving villagers. Much of the world outside of the U.S. remains a realm not unlike a Mad Max movie. And the creditors know it.

People who take the psychogenic fugue approach to debt repayment should probably choose more American-friendly environments such as Cancun, Jamaica, or Costa Rica as their destinations. Think of it this way. You may get malaria, but you won't freeze to death. And you can always ward off malaria with a daily G&T - the tonic water contains quinine.

Point XII: The Federal Reserve is necessary but vastly overrated. After skimming through Alan Greenspan's 2007 book, The Age of Turbulence, I arrived at this conclusion. For you conspiracy buffs and Ron Paul groupies, pay attention. The Federal Reserve is run by a board of seven trustees called governors, one of whom is appointed to an eight-year term as chairman by the President. The Fed has two main functions.

First, it runs the FDIC (Federal Deposit Insurance Corporation), which insures individual bank accounts up to $100,000 in the event of a bank failure. Second, the Fed sets monetary policy by adjusting interest rates by purchasing and selling government bonds and T bills.

That's all.

Much to the chagrin of the conspiracy crowd, the Fed is not run by a sinister cabal of OPEC ministers, the Bilderbergers, the Trilateral Commission, the Orthodox Union, Jews for Jesus, or the Free Masons. None of the above-mentioned groups has offered me an advance on this book, and since I have no intention of joining their ranks, I have no reason to cover for them. Ain't logic a bitch?

Point XIII - Economics is an inherently unpredictable science. In fact, it is hard to justify calling it a "science" as much as a sophisticated game of chance. That's why economists couch their predictions in statements that begin with "other things being equal" or "if current trends continue." These people do not have crystal balls at their disposal. Not even Jim Cramer on Mad Money, much as I love the man's shtick. Their guesses, albeit educated ones, about the stock market, tend to be as accurate as information from a Tarot card reader or Psychic Friends Network. The main difference is that a stockbroker usually charges more than the latter two services.

In conclusion, the realm of economics is more like a Las Vegas crapshoot than most people realize. Investing your hard-earned money in the global economy is tantamount to walking into the high roller area of a casino. On the one hand, nothing ventured, nothing gained. On the other hand, once you've entered the arena, you're in a realm summed up by Kenny Rogers in his hit song 'The Gambler.' "You gotta know when to hold 'em, know when to fold 'em, know when to walk away, know when to run." If you aren't into country music, keep in mind the one Jim Cramer adage worth repeating:

"Bulls make money. Bears make money. Hogs get slaughtered."

Now for some parting words of wisdom courtesy of Dr. Katz.

At the end of the day, people will spend their last dollar on food, energy, health care, and medication. The future is very bright for companies manufacturing fertilizer, transgenic seeds, liquefied natural gas, and pharmaceuticals. As the world's population continues to grow, these economic sectors will expand, and their stocks will rise.

Let's make a deal. Buy this book; invest some of your money in the above-mentioned stocks; email me back in thirty years, and let me know how it all turned out. By the year 2042, I should be collecting social security (if there's any money left in the general fund), and I'll be sure to respond with an emoticon of some sort:

It's the least I can do for my fans!

From Commander in Chief to the Imperial Presidency
The Growth of Presidential Power from 1865-2014

Evolution of the American Presidency

Today, Americans and people around the globe regard the President of the United States as the most powerful political figure in the U.S. (and arguably the world). Until the U.S. Civil War, however, Congress dominated the political landscape in Washington D.C. far more often than the President. This was more or less the intention of the Framers, who feared that without sufficient checks on his power, the President would become a de facto monarch. Abraham Lincoln's presidency marked an abrupt shift in the balance of power. Following Lincoln's election in 1860, the South seceded, and Congressional power all but collapsed.

At the time of Lincoln's assassination in 1865, few historians would have predicted that the importance of the presidency would recede relative to Congress for the next 35 years. Yet that is exactly what happened. In 1901, the assassination of William McKinley brought Theodore Roosevelt into the White House, and an unmistakable trend began. Successive Presidents either retained a degree of power comparable to that of T.R. or else expanded the role and influence of the Presidency even further. Presidential power peaked under Franklin Roosevelt, the only 4 term president in U.S. history. Although Congress passed the 22nd Amendment in 1947 limiting the President to 2 terms in office, FDR's successors have, with a few exceptions, pushed their agenda through Congress and not vice versa.

Between 1865 and 1930, strong Presidents in the tradition of Abraham Lincoln were the exception rather than the rule. The rise in Presidential prestige and influence, especially after 1930, was the result of several convergent historical trends: (1) The emergence of the U.S. as a global economic and military power by 1900; (2) The U.S. leading the Allies to victory in WWI and WWII; (3) The invention of ever faster forms of communications technology (telegraph □ radio TV internet); and (4) The President as the center of media attention, most notably since the time of FDR.

1865-1930

Reconstruction and the Election 1876: The Presidency in Retreat

Ulysses S. Grant took office in 1869 during the period of Radical Reconstruction. Northern Congressmen were determined to punish the South for the more than 400,000 Union casualties suffered during the Civil War. Radical Republicans would not recognize Congressional delegations from states that had seceded and were determined to make readmission to the Union a protracted ordeal for the South. To this end, they passed the 13th, 14th, and 15th Amendments and later impeached President Andrew Johnson. This was the first time a U.S. President had ever faced such humiliation. Congress intended to drive home the point that although Johnson had remained loyal to the Union, he was from Tennessee and his pro-Southern sympathies would not be tolerated. Johnson survived the impeachment trial by a single vote in the Senate and was not renominated by the Democratic Party in 1868.

Against this background, Grant began his term in office as a hugely popular war hero. Grant supported Reconstruction during his first term; in 1870, he signed the Force Acts, enabling state governments to use federal troops to crush the KKK and similar terrorist groups throughout the South. After 1872, however, violent white extremists in the South began to regroup. As Northern support for Reconstruction ebbed, Grant took half hearted measures in dealing with the growing Southern backlash against freed blacks and their Republican allies.

By 1876, Republicans had lost control of state governments in the South with the exceptions of Florida, Louisiana, and South Carolina. In the Presidential election held that year between Rutherford B. Hayes (a Republican) and Samuel Tilden (a Democrat) neither candidate won a majority of votes in the Electoral College. Because the contest was marred by numerous allegations and incidents of fraud on both sides, Congress assembled a bipartisan Electoral Commission to resolve the dispute. After secret negotiations, the Electoral Commission awarded the disputed electoral votes to Hayes. In return for his victory, however, Hayes agreed to withdraw all federal troops from the South, end Reconstruction and not seek reelection in 1880.

Teddy Roosevelt and the Progressive Era

In 1900, the Republican Party purposely chose Teddy Roosevelt to be William McKinley's running mate knowing that the vice president held almost no real power. They hoped that four years out of the spotlight would dampen the immense popularity T.R. enjoyed in the press for his exploits during the Spanish-American War. Their plan backfired when a deranged anarchist assassinated McKinley a few months after his inauguration in 1901. Much like Garfield's assassination in 1881, no one could have predicted this random act of violence. Conservative Republicans shuddered at the thought of T.R. in the Oval Office, but all they could do was watch as he implemented a Progressive agenda over the next 8 years.

Domestically, T.R. pushed legislation through Congress to weaken the power of monopolies, strengthen labor laws in favor of unions, and protect the public through government oversight of food processing and drug manufacturing. Landmark legislation passed during T.R.'s presidency included the creation of the Department of Labor and Commerce in 1903 and Pure Food and Drug Act of 1906. As a conservationist, T.R. oversaw the creation of the National Forest Service in 1905 and the expansion of the National Park system to include the Grand Canyon.

In the realm of foreign policy, T.R. was determined to project American power around the globe. Two events marked the ascendance of the U.S. in world affairs: the construction of the Panama Canal and the Cruise of the Great White Fleet from 1907-1909. In Panama, the French effort to build a canal connecting the Atlantic and Pacific Oceans was faltering due to poor planning, tropical disease, and opposition from the government of Colombia, which controlled Panama at the time. By 1903, France was on the brink of abandoning the project altogether. T.R. decided that Colombia's government was the major obstacle standing in the way of the canal's completion. Over the objections of his critics, T.R. supported Panamanian rebels as they launched an uprising against Colombian rule. In 1903, Panama broke away from Colombia, and Congress immediately recognized Panamanian independence. Colombia could no longer interfere with the construction of the Panama Canal, which was completed in 1914.

In the case of the Great White Fleet, T.R. decided that sending a force of 16 battleships on a three year voyage around the world would be an unprecedented display of U.S. military strength. When Senator Hale objected to the idea as a waste of money, T.R. pointed out that the fleet had already set sail and if he was so concerned about cost, the Senator should "try and get it [the fleet] back." (Military History, 2014) Of course, T.R. knew the public would be in an uproar if Congress failed to do everything possible to ensure the fleet's safe return. As with the Panama Canal, T.R. emerged triumphant; his critics could do little more than roll their eyes and grumble.

What made T.R.'s foreign policy accomplishments possible? Traditional answers focus on T.R.'s charisma and tremendous popularity but tend to ignore the deeper reasons: an explosion in U.S. industry (railroads, steel, and later electricity) from 1865 to 1900 combined with a massive influx of immigrants from Europe. To put it in perspective, the U.S. population on the eve of the Civil War was an estimated 31 million. By the time of the 1900 census, the U.S. population had more than doubled to 76 million people (U.S. Census Bureau, 2000).

Woodrow Wilson: World War I and the Isolationist Backlash

Wilson won the 1912 election when Teddy Roosevelt ran as an independent candidate with the Bull Moose Party. William Howard Taft, the incumbent, had accomplished little during his one term in office, and was secretly relieved by the outcome of the election. Taft became the only U.S. President to serve as chief justice on the U.S. Supreme Court, which he did from 1921 until his death in 1930.

Wilson watched with concern as war erupted in Europe in 1914 following the assassination of Archduke Franz Ferdinand in Sarajevo. The U.S. held the balance of military power in the Western Hemisphere, but joining forces with the British and French to defeat Germany and the Central Powers held little appeal for most Americans.

At the start of WWI, the United States was officially neutral but soon began shipping arms and supplies to the British. In response, the Germans set up an Atlantic naval blockade and engaged in unrestricted submarine warfare. Then in 1915, a German U-boat sank the luxury liner Lusitania, killing nearly 1,200 people, 128 of them Americans. Many Americans clamored for President Wilson to enter the war against Germany right then and there. Wilson bought time by warning Germany repeatedly to cease all submarine attacks on vessels flying the U.S. flag.

Tensions continued to simmer for the next two years as war hawks in Congress looked for any reason to declare war on Germany. The final straw came in the form of a diplomatic blunder known as the Zimmermann Telegram. In January 1917, British agents intercepted an encrypted telegram from Arthur Zimmermann, the foreign Secretary of the German Empire to von Eckhardt, the German minister to Mexico. The message stated that if Mexico attacked the U.S., Germany would one day help Mexico regain the territory it had ceded to the U.S. under the treaty of Guadaloupe-Hidalgo in 1847.

Predictably, much of the U.S. public was outraged when the press broke the story of Germany's plot to help Mexico conquer U.S. territory. In hindsight, this fiasco reflected Germany's growing desperation to avoid direct conflict with the U.S. In any event, less than a month after the telegram was intercepted, Germany resumed unrestricted submarine warfare, making a showdown with the U.S. inevitable. Congress declared war on Germany on April 6, 1917. Wilson justified American military intervention in Europe as the only way "To make the world safe for democracy". After 18 months of fighting and 100,000 American casualties, the U.S. led the Allies to victory. The Kaiser abdicated and fled to Belgium. Germany signed the Treaty of Versailles on November 11, 1918, ending WWI. Western Europe hailed President Wilson as the hero who had put an end to the Great War.

In 1919 came the isolationist backlash. In spite of Wilson's best efforts, the Senate did not ratify Treaty of Versailles nor did it allow the U.S. to join the League of Nations. Although this did not directly lead to the rise of fascism and the Axis Powers, it severely weakened the WWI Alliance. With the U.S. gripped by isolationism, Britain and France lacked the confidence to force a showdown with Germany in the mid-1930's, when decisive military action might have headed off a German assault on Europe. Instead, the U.S. waited until after Pearl Harbor to intervene militarily in Europe and Asia. By that time, Nazi Germany had captured France, blockaded Britain, and invaded the Soviet Union. A decade earlier, Japan used its foothold in Korea to gain control of Manchuria. As with Nazi Germany, the League of Nations condemned Japanese aggression but could do nothing to stop it. By 1941, Japanese troops had invaded eastern China. Immediately after attacking Pearl Harbor Japan invaded Hong Kong, the Philippines, and most of Southeast Asia.

1930 – Present

FDR was elected President in 1932, one of the worst years of the Great Depression. By the time of his death in 1945, the economy had roared back to life, and the U.S. was on the brink of victory in WWII. A coalition of Democrats held a solid majority in both houses of Congress from 1932 to 1952. Although no single factor accounts for FDR's overwhelming success as a president, it is difficult to exaggerate the importance of the New Deal. So called alphabet organizations like the TVA (Tennessee Valley Authority), CCC (Civilian Conservation Corps), and WPA (Works Progress Administration) provided jobs at a time when over 25% of America's workforce was unemployed. Other key pieces of New Deal legislation included Aid to Families with Dependent Children and the Social Security Administration, created in 1935. Without the New Deal's solid track record of job creation and economic recovery, it is doubtful whether FDR would have enjoyed enough popular support to run for a third term in 1940.

FDR was quick to adopt technology in order to spread his message to the widest possible audience. For example, he gave a series of weekly broadcasts called fireside chats on the radio. Given his immense popularity, the media devoted tremendous time and resources covering every aspect of FDR's public life. At the same time, FDR's private life was strictly off limits. Aside from a few family portraits, almost no pictures exist showing FDR in a wheelchair or wearing leg braces.

In the arena of foreign policy, FDR adopted a strategy similar to that of Wilson during WWI. He maintained neutrality as long as possible to buy time for a U.S. military buildup. In 1940, FDR campaigned for reelection and instituted the first peace time draft in U.S. history as war raged in Europe and Asia. FDR provided weapons and other supplies to the British, behind the scenes at first, and then more openly following the passage of the Lend-Lease Act in 1941.

In the wake of the Japanese attack on Pearl Harbor on December 7, 1941, isolationism evaporated, the U.S. mobilized the largest army in its history, and over the next four years defeated Nazi Germany and Imperial Japan. FDR was elected to a 4th term in office in 1944 but died in April 1945, a few weeks before Germany surrendered unconditionally. By the end of WWII, the U.S. emerged as a nuclear superpower; not surprisingly, the President came to be regarded as the most powerful leader on the planet.

The Nixon Era: Watergate and its aftermath

After serving as vice President for 8 years, Richard Nixon expected to succeed Eisenhower but lost to John F. Kennedy in the 1960 election. As Lyndon Johnson and the Democrats lost popularity over the war in Vietnam as well as nationwide riots following the assassination of Dr. Martin Luther King, Nixon knew his moment of opportunity had arrived. In 1968, Nixon was elected by a solid margin, promising a return to Law and Order in a nation spinning out of control. In the years that followed, Nixon created the EPA (Environmental Protection Agency), reestablished diplomatic relations with China, and ended the U.S. role in Vietnam on less than favorable terms for Saigon.

Then in 1972 the Watergate scandal broke. A group of five intruders calling themselves the Plumbers broke into the headquarters of the Democratic National Committee, ostensibly to steal confidential documents and install wiretaps. The intruders were arrested. When asked to comment, Nixon dismissed the break in as a "third rate burglary." Over the next two years, reporters Bob Woodward and Carl Bernstein of the Washington Post kept the story alive.

As evidence mounted that the President knew about the break in ahead of time then ordered a cover up, pressure mounted in Congress to draft articles of impeachment. Rather than face an impeachment trial, Nixon resigned in August 1974.

While historians continue to debate the long term ramifications of the Watergate scandal, its short term legacy consisted of two weak Presidents (Gerald Ford and Jimmy Carter) and the War Powers Act of 1973. Congress passed the War Powers Act in order to curtail unilateral military action by a President. Essentially, if the President orders U.S. troops to intervene in combat overseas, Congress reserves the right to order the President to withdraw those troops from the war zone after 60 days of hostilities. Interestingly, Congress has never once invoked the War Powers Act since 1973. For example, when Operation Iraqi Freedom lost popularity by 2006, the Democratic majority in Congress did not demand that George W. Bush justify an ongoing occupation of Iraq let alone force him to withdraw U.S. troops. All in all, Watergate seemed to have few tangible effects on the Presidency aside from the removal of hidden microphones and recording devices from the Oval Office after Nixon's resignation. In that sense, Presidential correspondence has become even less accessible to both Congress and the public since the time of Nixon.

Ronald Reagan: the Republican Revival and the Iran-Contra Affair

Ronald Reagan was elected in a landslide in November 1980. At that time, Americans were disappointed with Jimmy Carter and his abysmal track record at home and abroad: inflation had hit double digits by the late 1970's; a fundamentalist regime in Iran held American diplomats hostage for over a year; even more ominously, the Soviets invaded Afghanistan in 1979, possibly as a test run for a future invasion of the Middle East. Carter's responses were well intentioned but woefully inadequate. Rather than increase government spending to head off the strangulating effects of high inflation and high interest rates, Carter did little more than lecture the country on the need to endure tough times. Rather than invade Iran and topple a fanatical Islamic regime, Carter pursued endless rounds of fruitless negotiations. A poorly planned military attempt to rescue the hostages failed, making Carter look utterly inept as Commander in Chief. Finally, to protest the invasion of Afghanistan, Carter placed an embargo on grain sales to the Soviet Union and boycotted the 1980 summer Olympics in Moscow. Critics pointed out that these gestures were well meaning but essentially hollow, and the American public agreed.

During his first term, Reagan used a combination of rhetoric, diplomacy, and behind the scenes deals to undermine the global influence of the Soviet Union. In one famous speech, Reagan referred to the Soviet Union as "an evil empire" and to communism as "the focus of evil in the modern world." Against the advice of critics and fiscal conservatives, Reagan massively increased government spending, especially in the realm of military technology. He also found ways to funnel weapons to any resistance movement dedicated to fighting communist regimes in general and the Soviets in particular. In Afghanistan, the U.S. supplied weapons to resistance fighters called the mujahedeen via Pakistani, Saudi, and Israeli intermediaries. By 1989, the Soviets lost an estimated 13,000 troops occupying Afghanistan and withdrew their forces that same year (Taubman, NY Times, 1988). In Nicaragua, Reagan sought to arm anti-Communist guerillas called the Contras even though Congress had passed a law limiting assistance to Nicaraguan opposition groups to humanitarian aid. This set the stage for the Iran-Contra Affair, discussed next.

Reagan won reelection in a landslide in 1984. The challenger, former VP Walter Mondale, won his home state of Minnesota and Washington D.C. Unlike his first term as President, Reagan's second term was overshadowed by the Iran-Contra affair, which the media first reported in 1986. The Iran-Contra Affair is often described superficially as "trading arms for hostages". The Iranians released American and British hostages in exchange for weapons in their seemingly endless war against Iraq.

The twist was that Colonel Oliver North used profits from these secret weapons sales to buy additional weapons, which the Central Intelligence Agency (CIA) and various mercenaries delivered to the Contras. In order to escape detection by the Congressional Armed Services Committee, all funds used to arm the Contras were hidden in secret bank accounts in Switzerland and Brunei.

Upon deeper examination, the affair revealed the lengths high ranking U.S. military officers would go to bypass Congressional legislation, in effect breaking the law. Congress was no doubt surprised and perturbed that the military and CIA could flout the law with relative ease via a global network of secret agents, weapons smugglers, and overseas bank accounts. The outcome of the Iran-Contra Affair was anticlimactic to say the least. Admiral Poindexter resigned from the Navy, and Col. North was fined $150,000 and sentenced to 1,200 hours of community service (Walsh, Iran / Contra Report).

Reagan may or may not have known about the Iran-Contra affair, but VP Bush, as a former CIA director, was almost certainly in the loop and may even have helped orchestrate the affair. Most analysts think that Col. North and his boss Admiral Poindexter could not have set up an international network of arms dealers and secret bank accounts without CIA assistance. Although President Reagan was never directly implicated in the Iran-Contra Affair, it paralyzed his last two years in office.

Similar outcomes have followed every Presidential scandal since 1986. Soon after George H.W. Bush took office in 1989, he testified under oath that he had no direct knowledge of or involvement in the Iran-Contra Affair. When critics dismissed his testimony as a pack of lies,

Bush simply responded that he had sworn under oath to tell the truth and therefore had not lied. Although this amounted to a circular argument, Congress never pursued the matter further.

Bill Clinton's impeachment in 1999 seems like an exception to this trend, but closer analysis reveals it was little more than a year long episode of political theater. Although the votes for and against impeachment were cast almost exclusively along party lines, it was apparent that few members of Congress considered an extramarital affair a "high crime or misdemeanor" worthy of removal from office. Unlike Andrew Johnson, Clinton survived impeachment by a margin of 17 votes.

Conclusions:

Today, there is little doubt that the Executive Branch receives more attention than any other branch of the U.S. government. Whether they admire or dislike the President as an individual, Americans generally assume that the President is not only the most powerful member of the Federal Government but also the most powerful leader on the world stage. Millions of people watch the State of the Union Address every year. Millions drop whatever they are doing to watch a Presidential press conference; an estimated 29 million people follow President Obama on Twitter; and an additional 42 million have liked his Facebook page (FB.com, 2014).

Much to the chagrin of Congress, almost no one watches C-Span. On a fundamental level, the President is the star of the show in U.S. politics. These are some of the key events that led to this outcome:

•   Grant implemented most of the Republican Party's agenda for Reconstruction, although his two administrations were plagued by corruption and incompetence. The disputed election of 1876 paved the way for the end of Reconstruction and a weakened presidency for the next 25 years.

•   Theodore Roosevelt set the tone for the Imperial Presidency with activist legislation, the Panama Canal project, and the cruise of the Great White Fleet. It was clear throughout his presidency that T.R and not Congress was in the driver's seat.

•   Wilson and FDR used similar strategies to lead the U.S. to victory in World War I and World War II, respectively. Because world history unfolded very differently after 1945 compared to 1918, the foreign policy legacies of these two presidents stand in stark contrast to one another.

•   Richard Nixon became the only U.S. President to resign the office of the Presidency. Nonetheless, Watergate will probably go down as an historical anomaly. In the years since Watergate, the media's power has declined dramatically even as the President's has surged.

•   Ronald Reagan regained any Presidential power lost in the wake of Nixon's resignation. The Iran-Contra affair paralyzed his last two years in office, but, unlike Watergate, there was virtually no chance of Reagan resigning let alone facing impeachment.

Essay 112: Election 2016

This book would not be complete without an essay devoted to the Presidential Election of 2016. It's hard to know where to begin when discussing this Presidential Election. Suffice it to say, future historians will be astounded that in a country of 312 million people, the two finalists for the Presidency were former First Lady Hillary Clinton and media sensation Donald Trump.

The case for Clinton / her true motives – Although the media has taken up the narrative that Hillary Clinton is running to make history as the first female President of the U.S., her motives are abundantly clear – and superficial - to anyone old enough to remember Bill Clinton's presidency. Here's a quick recap for millenials: Bill Clinton unseated George Bush Sr. in 1992, and for 2 years the Democratic Party held the White house and a majority in both houses of Congress. Nonetheless, they accomplished nothing to speak of between Jan. 1993 and the midterm elections of 1994. When the GOP won the House and Senate for the next 12 years, the Democrats desperately looked for a scapegoat. Hillary Clinton quickly became the lightning rod for both parties as Bill Clinton's presidency slowly unraveled.

When the Republicans impeached Bill Clinton in 1999, few people were surprised that Hillary did not divorce him or at least move out of the White House. Her reason was obvious: By riding out the political firestorm alongside her husband, Hillary was in effect saying "I am entitled to become President some day because I was a Democratic martyr during the 1990's."

The case against Hillary: Forget the deleted emails and the corruption surrounding the Clinton foundation. Far and away, the scandal that should have torpedoed Hillary's campaign was the 4 Americans savagely murdered by terrorists in Benghazi, Libya.

The case for Trump is more straightforward: Trump's slogan is Make America Great Again. Although this appeals to many older voters and Republicans, this message does not resonate with young / naïve people or with many minorities who feel they are treated as second class citizens regardless of who occupies the White House.

The case against Trump – he presided over a few casinos that went bankrupt. Shrug. Second argument - he lacks self control around beautiful women. Ok…so did JFK, LBJ, FDR and half the other Presidents. Once again, shrug.

Conclusion: Liberals, Democrats and Hillary supporters are not the adults in the room; the sooner they get swept out of national politics the better.

The lame stream media will act genuinely shocked if (or rather when) Trump wins. The reason is also glaringly obvious. The media remains willfully oblivious to the elephant in the room – the majority of America distrusts the media. We agree with Trump's statement – the media became a biased bunch of corrupt, dishonest dirt bags a long, long time ago. Trump is the only candidate with the courage to say what no one in the media has the backbone admit – the emperor has no clothes. Since 2012, millions of Americans have taken this message to heart. And this November, I believe the American electorate will prove Trump correct. The party is over for establishment Democrats and Republicans alike. It's time to clean house with a flame thrower, and Trump was the spark who ignited the fire. God bless him for that!

Prediction – Trump wins with 280 Electoral votes vs. Clinton 258 ☐

Just before Inauguration Day later I posted this on an online forum called Quora:

No one factor torpedoed Hillary's campaign; rather her defeat was a case of death by 1000 paper cuts.

For starters, Hillary started her presidential bid with a high negativity rating among likely voters. Some voters never forgave her inept reaction to the Benghazi attacks in 2012. Others disliked Bill Clinton and lumped Hillary together with the former President instead of judging her on her own record. Yet other Democratic voters favored Bernie Sanders and lost their enthusiasm for the Democratic Party once Sanders dropped out of the race. Finally, there was the simmering email scandal which Clinton tried her best to ignore but the media never put to rest. Keep in mind that these events eroded Clinton's support base even before she was formally nominated at the Democratic convention.

Once the conventions took place, the Republicans (especially Trump supporters) swung into action. After securing the GOP nomination, Trump effectively moved the arena of the Presidential race from television to online social media. A glance at Clinton's Facebook page showed 5.5 million likes while Trump's page sported over 12 million likes. Although most pundits dismissed this statistic as irrelevant, it underscored the extent to which the media was ignoring large blocs of Trump voters. An even more worrisome sign (again largely dismissed by the media) was the size of the crowds at each candidate's rallies. Trump rallies consistently drew 3,000 or more supporters. In contrast, most Clinton rallies could be held in a high school gymnasium with room to spare.

By the time the debates took place, it is safe to assume most voters had their minds made up. While neither candidate scored a crushing victory in the 3 debates, Trump's rhetoric was still more memorable. Two remarks stand out. First, Trump acknowledged that Clinton had years of experience as a Senator and Secretary of State but then portrayed this as "bad experience" - evidenced by the Benghazi scandal, a resurgent Russia and 33,000 deleted emails. At the final debate, Trump maneuvered Clinton into her most memorable gaffe. Clinton commented "It's a good thing people like Donald Trump aren't in charge of enforcing the laws in the United States." A split second later came Trump's retort, "Yeah, because you'd be in jail." The audience erupted into applause. And Clinton's campaign never recovered.

A few days before November 8th, the national media still claimed Clinton held a 3 to 5 point lead in most polls, but more sophisticated pollsters knew otherwise. As the results came in, most states fell predictably into the Trump or Clinton columns. The telltale sign of a Trump victory came as Florida, North Carolina and Ohio went to Trump. Soon afterwards, exit polls showed Trump in the lead in Wisconsin, Michigan and Pennsylvania. These were states the Democrats won for the past 7 Presidential elections - and not surprisingly, they took this "Blue Wall" for granted. Sure enough, a surge of disgruntled voters in all 3 states tipped the results to Trump, and he won the 2016 election.

Essay 113: U.S. Civil War - The Battle of Shiloh

The Battle of Shiloh was fought on April 6-7, 1862, nearly a year after the U.S. Civil War erupted. The battle took place at Pittsburg, landing in western Tennessee. Those two days of fighting saw the heaviest casualties of the war up to that point. Historians estimate that Union and Confederate forces altogether suffered nearly 24,000 casualties at Shiloh. As devastating as the battle was, Shiloh was a prelude to 3 more years of carnage, which ultimately claimed over 800,000 American lives. This article will explore the battle of Shiloh and its significance on the outcome of the Us.S. Civil War.

The Battle

On April 6, 1862, the Battle of Shiloh began as the Confederate Army of the Mississippi under General Albert E. Johnston launched a pre-dawn attack on the Army of Tennessee's lightly fortified positions near Pittsburg landing. The Army of Tennessee was commanded by General William T. Sherman and Gen. Ulysses S. Grant, two of the Union's most able commanders. Grant was several miles away when the attack began, awaiting the arrival of Gen. Don Carlos Buell and 17,000 reinforcements. Sherman's troops, numbering about 40,000, felt the full force of the Confederate army's 44,000-man assault.

Supposedly, the Confederate plan was to break the Union's line, push most of the divided force east into the Tennessee River, and trap any survivors in a swamp northwest of Pittsburg landing. Although Johnston's attack caught the Union army off guard, most troops on both sides had never experienced combat before. The battle soon disintegrated into a brawl; in the melee that followed, the Union took the brunt of the day's casualties. Still, the Confederate plan to force a Union retreat failed.

The second day of the battle was a different story altogether. Johnston was killed in action on the first day at Shiloh. His second in command, Gen. P.G.T. Beauregard, was unaware that Buell's forces had arrived in the middle of the night. In contrast to the inexperienced troops the Confederates faced on the first day, most of Buell's troops had fought at the Battles of Fort Henry and Ft. Donnellson. They also brought a large supply of artillery and ammunition with them.

As the battle raged, Union forces slowly pushed the Confederate lines back to their positions at the start of the battle. By that afternoon, Beauregard ordered his troops to retreat southward into Mississippi. Union troops did not pursue them until the following day.

Consequences for the Confederacy: The South never smiled after Shiloh.

Although neither side could claim a decisive victory at Shiloh, most historians point out that the South suffered more harm in the long run. Although the Union advanced through Tennessee ground to a halt, General Beauregard's forces failed to recapture any Union-held territory. Second, the loss of Albert Johnston deprived the South of a military leader with leadership ability and experience on par with General Lee. Finally, the Confederate casualties at Shiloh topped 10,000. This was somewhat lower than the Union's casualties, but unlike the Northern armies, the troops of the Southern armies could not be replaced. By 1863, the Confederate war effort had all but collapsed in the West.

Consequences for the Union

In the aftermath of Shiloh, General Grant sent Sherman in pursuit of the Confederate forces. Sherman spent most of the next two years fighting in the

West. In 1864, Sherman would advance through Tennessee and ultimately capture Atlanta, Georgia. Grant himself was transferred to the Western theater, where he would stay through the capture of Vicksburg, Mississippi, in July of 1863. With the surrender of Vicksburg, Union forces took complete control of the Mississippi River, effectively winning the war in the West.

Impact of Shiloh on the overall course of the Civil War

In the year following Shiloh, some of the bloodiest battles of the war unfolded in the East: Antietam, Fredericksburg, Chancellorsville, and finally, Gettysburg, which, in retrospect, marked the beginning of the end for the Confederacy. By 1864, Grant had become adept at siege warfare and employed these tactics in an ultimately successful war of attrition against Robert E. Lee and the Army of Northern Virginia.

Ultimately, no one can say for certain if the decision to send Grant out West prolonged the Civil War by over a year or, conversely, if the experience he gained at Vicksburg was instrumental in defeating General Lee. As for Sherman, his march from Atlanta to the sea and then north through the Carolinas brought the specter of total war to much of the Confederacy. What convinced Sherman to adopt this strategy remains a matter of debate; however, there is no doubt about the outcome. In April of 1865, General Lee surrendered to Grant at Appomattox, and the end of the war was at hand.

World History & International Politics

Essay 114: The Dreyfus Affair

This notorious scandal, known in France as L' Affaire, began in 1894, with the arrest and court martial of Captain Alfred Dreyfus on charges of committing espionage on behalf of Germany. By the time the scandal ended in 1906, France's Third Republic had been shaken to its foundations. Not only did the Dreyfus Affair expose the entrenched corruption within the ranks of the French military, but it also revealed the pervasive Anti-Semitism latent in many segments of French society a century after the French Revolution and its slogan of Liberty, Equality, and Fraternity. The bitter political and social divisions that surfaced during the Dreyfus Affair haunted France for decades afterward. Ultimately, the Dreyfus Affair had unpredictable repercussions at the time but, in all likelihood, changed the course of modern history.

Alfred Dreyfus was born in 1859 into a Jewish family living in the French province of Alsace. Around the time of the Franco-Prussian War in 1871, the Dreyfus family moved to Paris. After studying at the Ecole Polytechnique, Dreyfus entered the French army and was promoted to the rank of captain in 1889.

Five years later, letters allegedly containing French military secrets turned up in a waste basket in the German embassy in Paris. The true culprit, Major Ferdinand Esterhazy, named Dreyfus as the guilty party, claiming his handwriting matched that on the recovered letters. The military tribunal that tried Dreyfus also denied him the right to view the evidence being used against him. Some historians point out that France's defeat in 1871 and the subsequent loss of Alsace-Lorraine to Germany made the French military establishment more inclined to believe the fabricated tale of Dreyfus's disloyalty.

At any rate, considering that Dreyfus was a Jew and from a territory now controlled by France's archenemy, the trial's outcome was a foregone conclusion. In 1895, Dreyfus was found guilty and sentenced to life in prison on Devil's Island, which was tantamount to a death sentence - 4 out of 5 people sentenced to prison on Devil's Island never returned alive. After Dreyfus was sent to Devil's Island, most people rapidly lost interest in the case.

In 1896, however, rumors began to surface that the military establishment had framed Dreyfus rather than expose a higher-ranking officer as the traitor.

Major Georges Picquart investigated the trail of evidence used to convict Dreyfus and concluded that the espionage charges were baseless. When Picquart's commanding officers ordered him to remain silent about his findings, the infuriated major spilled the story to Emile Zola, one of France's most prominent writers.

In 1898, Zola published these scathing indictments in a work called J'Accuse ("I Accuse"). In response, the military transferred Picquart to Tunisia, and Zola himself fled to London rather than face imprisonment on charges of libel. Nevertheless, Zola's persuasive work had its desired effect. Public opinion turned against the French military; pressure mounted on the French government to reopen the case, and in 1899, Dreyfus was brought back to Paris for a second trial.

By this time, France was divided into pro- and anti-Dreyfus factions. The circumstances surrounding the retrial were bizarre, to say the least. The military's file on Dreyfus had grown during his years on Devil's Island. Additional letters had appeared, ostensibly strengthening the charges against him. Soon after this "new evidence" came to light, Major Hubert Henry admitted to forging these additional letters and subsequently committed suicide.

Nonetheless, the military tribunal again found Dreyfus guilty. At this point, France's President Emile Loubet interceded and pardoned Dreyfus. In 1906, the French military reinstated Dreyfus and later promoted him to the rank of Major. After serving in WWI, Dreyfus retired and spent the rest of his life in France, dying in 1935 at the age of 75. Major Esterhazy left the French military in 1898 and spent the rest of his life in England, where he died in 1923.

Consequences of the Dreyfus Affair

1) Calls for the complete separation of Church and State in the Third Republic intensified; an official policy of secularism went into effect in 1905.

2) Public distrust of the French military would plague France through WWI and beyond. Also, the Dreyfus Affair revealed how polarized France's citizens had become in terms of their support for the government of the Third Republic. Progressives, anti-militarists, and socialists were no doubt pleased with their government's actions; on the other hand, the military and Catholic church became the Third Republic's staunchest opponents.

3) Theodore Herzl, a journalist for a Vienna newspaper, covered Dreyfus's first trial in 1895. When he witnessed a mob in Paris chanting "Death to the Jews," he concluded that the Jewish people had no future in Europe and needed to return to their historic homeland in Israel, a province of the faltering Ottoman Empire. Herzl galvanized support for political Zionism as the movement came to be known. In 1897, at the First Zionist Congress in Basel, Switzerland, he prophetically remarked that a Jewish state would be created 50 years in the future. True to Herzl's words, in November 1947, the United Nations partitioned the remainder of the British Mandate over Palestine, and the modern State of Israel declared its independence in May 1948.

Essay 115: Geography and International Relations

Archaeological records as well as written historical accounts provide overwhelming evidence that geography plays a pivotal role in international relations at the level of the nation-state. Exploring world history through the lens of geography reveals common themes applicable to ancient and modern times. This analysis will cover several states in the ancient and modern world, comparing and contrasting the influence of geography on their historical trajectories and international relations.

Ancient States: Mesopotamia and Egypt

Most scholars consider the Sumerian city-states of Mesopotamia to be the world's earliest known civilization. These city-states arose between the Tigris and Euphrates Rivers between 5,000 and 6,000 years ago. This was surely not a coincidence considering that Mesopotamia and the Fertile Crescent contain the oldest known evidence of human agriculture and animal domestication as well. Sumerian city-states gave way to empires as food production and metallurgy advanced (Diamond, 2003).

No significant geographical barriers meant that trade networks connecting Mesopotamia to Turkey and Egypt in the West and Persia in the East sprang up as early as 5,000 years ago. Mesopotamians also had access to the Persian Gulf, allowing for the possibility of trade with the Arabian Peninsula and India. In 1977, Thor Heyerdahl successfully sailed a ship called the Tigris from Iraq to the coast of India and finally to the Horn of Africa to demonstrate the feasibility of such a trade network (Best, 1984).

The downside, of course, was that, with the rise of horse drawn chariots and iron weapons, any region of Mesopotamia could conquer or be conquered by a large army of warriors mounted on horseback and equipped with advanced weaponry. In ancient times, a long succession of empires dominated much of Mesopotamia, including Akkad, Babylonia, the Hittites, Assyria, Chaldea, and Persia (Potok, 1978).

Alexander the Great (4th century BC) and later the Roman Empire were able to conquer so much of the Near East due to the absence of significant geographical barriers. In the 7th century

AD, the large scale use of horses and camels allowed Arab armies to crush the Byzantines and Persians in less than a decade. By 732, the Arabs controlled an empire stretching from Spain to India (Potok, 1978). During the 12th and 13th centuries, the region was ravaged by European Crusaders and Mongols, respectively. By the 1500's Ottoman Turks gained control of North Africa, the Near East, the Balkans, and most of the coastal territory of the Arabian Peninsula.

The decline of the Ottoman Empire and its collapse during WWI set the stage for British and French colonial governments to partition the region. These artificial borders (as opposed to specific geographical features) contributed greatly to the political instability of the Middle East after WWII. Nonetheless, geography continues to exert a powerful influence on economic ties and international relations in the Middle East today. Regional and international conflicts have erupted over control of the Suez Canal (1956); Israel's access to the Red Sea (1967); the Iran-Iraq war (which led to U.S. intervention to protect oil tankers in the Persian Gulf from 1985-1989); and Iraq's invasion of Kuwait, culminating in the Persian Gulf War of 1991.

Egypt traded with the Mediterranean, Mesopotamia, and East Africa. More insulated geographically than Mesopotamia; this contributed to long periods of political stability.

Except for an invasion by the Hyksos (18th century BC) and occasional conflicts with the Hittites and Sea peoples, ancient Egypt was largely buffered from the turmoil that frequently engulfed Mesopotamia. After 1000 BC, however, Egypt lost much of its military advantage compared to its neighbors.

Even its insular geography no longer posed a serious barrier to successive waves of invaders; Libyans, Nubians, Assyrians, and Persians all conquered Egypt between 900 and 500 BC (Potok, 1978).

India and China – virtual isolation within a continent

China is one of a handful of Old World areas that discovered agriculture independently. The beginnings of agriculture in China are less well characterized than for Mesopotamia but are thought to have started only a few centuries later. China became politically unified in 221 BC by Emperor Shi Hwang Ti, whose tomb contained the famous Terra Cotta Army. This emperor also started construction of defensive walls on China's northern border to discourage raids by Mongolian nomads. Centuries later, these earthworks would be connected into the Great Wall of China. Later emperors turned their attention to uniting China internally. By the early 7th century AD, the Chinese completed the Grand Canal linking the Yellow and Yangtze Rivers (National Geographic, 2013)

Despite intermittent civil wars and a 13th century Mongol invasion, China remained a single political entity for most of those 2,200 years. Throughout the medieval period, direct contact between Europe and China was limited to caravans traveling the Silk Road and Marco Polo's visit to the court of Kublai Khan in the late 13th century. During the Ming Dynasty (1368-1644), Chinese treasure fleets reached the Persian Gulf and perhaps the coast of East Africa.

Starting in the late 1400's however, Ming emperors ended the treasure fleets, dismantled the shipyards, and effectively ended contact between China and the outside world (Diamond, 2003).

India has had a far different historical course than China. The cities of Mohenjo Daro and Harrapa in the Indus Valley mark the emergence of an agricultural civilization in India circa 2000 BC. Several centuries later, the Aryan invasion through the Khyber Pass established the caste system. Two regional empires, the Maurya and Gupta, flourished between 500 BC and 500 AD; however disunity prevailed more often than centralized government in ancient and medieval India. Finally, by the late 1500's Akbar the Great united most of India under the Mogul Empire (Lal, 2001).

Throughout its history, India had cultural and economic contact with China, but much of this took the form of maritime trade; overland travel was infrequent. For example, in the 7th century AD, a Buddhist monk named Hsuan Tsang made a lone visit from China across the Himalayas to India. Eight centuries later, this story was published in an epic work called 'Journey to the West'; this book became a literary sensation in late medieval China, attesting to the rarity of long distance travel for so much of human history (Best, 1984).

A daunting geographical feature explains why these two populous states developed in near isolation from one another: the Himalayas are the world's highest mountain range. No army in recorded history has crossed them to invade India from China or vice versa. Even China's invasion of Tibet did not take place until 1951. South of China, the jungles of Southeast Asia and the lowland swamps of Myanmar and Bangladesh make travel between these two parts of Asia slow and difficult.

Modern states

U.S.A. From the time of the American Revolution through America's centennial in 1876, the U.S. expanded westward from a series of 13 colonies on the Eastern seaboard to a nation stretching from the Atlantic to Pacific Ocean. During this time, the population of the U.S. surged from an estimated 3 million people in 1776 to 46 million a century later (Doerner, 2014).

Meanwhile, the Native American population dwindled, largely as a result of epidemics, forced relocations (e.g. the Trail of Tears in 1830) and direct attacks by the U.S. Army as well as massacres by private citizens.

Somewhere between 250,000 and 500,000 American Indians lived in the continental U.S. at the time the Declaration of Independence was signed; this population shrank to 100,000 circa 1900.

Although geographic features like the Rocky Mts. and Mojave Desert slowed the pace of westward expansion from the Mississippi River Valley to the Pacific coast, no geographic barrier was formidable enough to halt the spread of European immigrants throughout the lower 48 states (Oliver, N.D.)

Beyond its domestic history, geography has shaped U.S. relations with Canada and Mexico. American relations with Canada improved with a series of political treaties following the War of 1812; however, the Great lakes and Canadian prairies acted as natural boundaries to limit northward U.S. expansion. After the resolution of a political dispute over the northwestern border of the Oregon Territory in 1846 (the so called 54o, 40' or fight controversy), the U.S. and Canada have never come close to military conflict in over 150 years (U.S. History, 2014).

On its southern border, the U.S. fought Mexico in 1848, capturing the territories of New Mexico, Arizona, and California. U.S. forces also captured Veracruz and Mexico City; nevertheless, the U.S. government chose not to annex any territory south of the Sonoran Desert. In 1853, the Gadsden Purchase marked the end of U.S. territorial expansion in the Southwest.

Globally, the U.S. began to project its military power in the decades after the Civil War (1861-1865). Mimicking the European bid for empire, the U.S. fought the Spanish-American War in 1898, capturing Cuba, Puerto Rico, Guam, and the Philippines in quick succession. Throughout the 20th century, U.S. foreign policy in the Western Hemisphere consisted of a mixture of the Monroe Doctrine (which prohibited further European colonies in the Americas); the Big Stick approach of Theodore Roosevelt, and the Good Neighbor Policy of Franklin Roosevelt. During the Cold War, the U.S. and Canada set up a defense alert network called the Distant Early Warning Line in the event of a Soviet invasion through Alaska. Aside from this unlikely scenario, the odds of a conventional land invasion of the U.S. via Canada or Mexico remain quite remote.

Japan – the most isolated economic power in the modern world.

Much of Japan's isolationism stems from its long history of tenuous contacts with East Asia and the rest of the world. This, in turn, is a direct result of Japan's remote geographic location. According to archaeologists, Japan was inhabited over 10,000 years ago by an aboriginal group known as the Ainu. Sometime around 400 BC, farmers from Korea migrated to Japan, pushing the Ainu northward to the island of Hokkaido. An ancient Chinese document dating to 100 A.D., contains the first historical reference to Japan, which the Chinese called Wa. Over 5 centuries later, Prince Shotoku Taishi traveled to China. Upon his return home, he introduced a Confucian style bureaucracy as well as Buddhism to Japan.

For the next thousand years, Japan developed a unique system of government in which the Emperor was relegated to figurehead status but simultaneously revered as a core element of the Shinto religion. Meanwhile, rival clans, headed by the Fujiwara and other families, battled each other for centuries on end. Political unity in the form of a shogun (basically a warlord who vowed allegiance to the emperor) came in 1185 when Minamoto Yoritomo established the first centralized Japanese government at Kamakura. For the next 4 centuries, Japan's involvement with the outside world was limited to repelling a Mongol invasion fleet in 1281, a chance encounter with a Portuguese ship in 1543, and an unsuccessful invasion of Korea in 1590 (Diamond, 2003).

For much of the Tokugawa Shogunate, which lasted from 1603 – 1868, the shoguns prohibited foreigners (including missionaries) from settling on the main islands of Japan.

They also forbade Japanese citizens from traveling abroad. Contact with the outside world was limited to a Dutch trading post in Nagasaki harbor as well as a few Japanese trading outposts on the coast of China (Diamond, 2005).

In 1853, Admiral Perry's fleet sailed into Tokyo harbor and forced Japan to open its ports to the outside world. Over the next few decades, Japan rapidly modernized its industries and military. By 1895, Japan fought a brief but decisive war with China and captured the island of Taiwan in the process.

In 1905, with American support, Japan defeated Russia in the Russo-Japanese War. In 1910, Japan annexed Korea and occupied the peninsula until the end of WWII in 1945. Since WWII, Japan has made strides to reconcile with Asia and connect with the wider world, most notably in the fields of science and international philanthropy.

New Guinea and Australia – islands at the opposite extremes of the geographical spectrum

Although humans reached Australia and New Guinea over 40,000 years ago, their radically different geographies had a profound impact on their subsequent histories. Owing to this island's remote location and rugged terrain, New Guinea societies became extremely fragmented. One legacy of this perpetual disunity is that the peoples of modern day New Guinea speak 1,000 languages, unrelated to any language groups in the outside world and often as different from each other as English and Chinese. The reason is straightforward: high mountain ranges, lowland swamps, and jungle made cultural exchanges, let alone political unity, impossible in New Guinea (Diamond, 1992).

Europeans reached the coast of New Guinea by 1526 but did not establish any permanent colonies until 1880. A combination of tropical disease and impenetrable jungle conspired to limit European exploration to a snail's pace. Only in 1930 did two gold prospectors, Leahy and Dwyer, reach the Eastern New Guinea Highlands, where they stumbled across a farming tribe of several thousand people. In 1938, the Archbald expedition encountered another uncontacted tribe of 50,000 natives, known as the Forë (Diamond, 1992). Papua New Guinea gained its independence from Australia in 1974. Its 2012 population was estimated at 7.2 million, nearly all of whom are natives. Europeans and other newcomers make up no more than 1% of the population (PNG, 2014).

Australia was colonized by Britain in 1788 and became a penal colony soon thereafter. Aborigines were killed in epidemics and direct assaults; by 1921, Australia's white population was 5.4 million, but its Aboriginal population had declined by 80% from an estimated 300,000 in 1788 to 60,000 (CCA, 2011). Unlike New Guinea, no significant geographic barriers or endemic diseases existed to prevent Europeans from gaining a rapid foothold in Australia.

Aside from the interior of Australia, which consists of mostly desert, European settlers occupied all arable land on the continent less than a century after Captain Cook's landing (Diamond, 2003).

Today, Australia enjoys modern technology and an urban lifestyle, with 60% of its citizens living in coastal cities. Although it maintains strong ties to Great Britain and remains part of the ANZUS pact with New Zealand and the U.S., the vast distance separating Australian cites from each other and the entire continent from the rest of the world represents an ongoing challenge for this country both economically as well as in the realm of international relations (Diamond, 2005).

Old World vs. New World Historical Trajectories

Civilizations arising in continental areas with few geographical boundaries established trade networks with their neighbors early on. As mentioned previously, Sumer, the Persian Gulf and the west coast of India had already established trade contacts 3,000 years ago. By the time of the Roman Empire, the East-West axis of the Old World including the Mediterranean, Middle East, India, and China were linked by trade routes (Diamond, 2003).

In contrast, while the Aztec and Inca Empires dominated their respective regions of the New World by the 15th century A.D., they never developed diplomatic relations with Native American chiefdoms 700 miles to the north (in the case of the Aztecs) or between the two empires themselves, which lay some 1,200 miles apart.

Several reasons account for the lack of long distance communication in the New World. First, until the arrival of Europeans, natives of the New World had no domesticated animals besides dogs, guinea pigs, and llamas. While llamas are useful as pack animals, they are still too small to carry an adult rider. A second, less obvious reason is the geographical layout of North and South America. Unlike the Old World, whose major axis runs East to West, the New World's major axis runs North to South.

Tenuous trade networks existed within and between various regions of the New World in the centuries before Columbus; however, several geographic barriers prevented any significant progress toward unification. In North America, the Great Plains and deserts of the Southwest blocked direct communication between Mexico and the Eastern U.S. Rainforests stretching throughout Central America and the Isthmus of Panama made overland travel extremely slow, even for the early Spanish explorers who had horses and ocean going ships. To put it in perspective, Balboa marched across Panama in 1513, but almost 20 more years would pass until Pizarro's conquest of the Inca Empire (Diamond, 2003).

In South America, the Andes Mountains and Amazon rainforest effectively divide that continent into two disconnected realms, at least at tropical latitudes. Even today, virtually all travel from the Atlantic coast of Brazil to the Pacific coast of Peru involves ships or aircraft. The Amazon is the second longest river in the world, but to this day, no one has built a bridge spanning it.

Few other places in the Old World, aside from the Sahara desert in Africa and the Himalayas in Asia, present such formidable geographical obstacles to travel.

Semi-isolated regions of Asia (India and China) eventually became linked to the rest of the Old World, but, as discussed, their respective histories were very different as a direct result of variations in regional geography. Island nations tend to harbor an isolationist world view corresponding to their own degree of geographic isolation; however, there is a vast spectrum in the realm of island nations and their role in modern day international relations, e.g. Japan vs. New Guinea.

The events and inventions of the 20th century – WWII, real time communication and travel technology – have altered certain aspects of international relations almost beyond recognition. Nonetheless, the legacy of geography continues to impact international relations on a profound level.

Essay 116: German Military Blunders of WWI

World War I (1914-1918) pitted the Central Powers, led by Germany, against the Triple Entente of Great Britain, France, and Russia. As the war unfolded, it became clear that while both sides possessed weapons far superior to any used during the 19th century, the ground maneuvers and logistics had barely changed since the U.S. Civil War, fought half a century earlier. Historians generally agree that the stage is set for a monumental catastrophe when military tactics lag far behind weaponry.

Sure enough, within a few months of the war's outbreak, both sides became locked in a bloody stalemate along the Western Front and, to a lesser extent, on the Eastern Front as well. Although no single decision was responsible for the quagmire, Germany's Schlieffen Plan, developed a decade before WWI, played a key role. Ultimately, Germany's decision to fight a two-front war, the sinking of the Lusitania, and an incident called the Zimmermann telegram all contributed to its defeat in WWI.

The Schlieffen Plan

In the aftermath of the Franco-Prussian War (1870-1871), Germany's main rivals in continental Europe were France and, as a distant second, Czarist Russia. At the dawn of the 20th century, the German military was no doubt confident it could defeat any individual European power in a land war. However, winning a two-front war could prove difficult even for the powerful German war machine. In 1905, a general named Alfred von Schlieffen anticipated this scenario and devised a strategy to maximize Germany's chances of victory.

Schlieffen's plan called for German forces to strike France, capture Paris, and force the French government to sign a ceasefire, all within a span of 6 weeks. With France effectively sidelined, Germany could withdraw its troops from the Western front and attack Russia, crushing the Czar's army before it could mount an attack on East Prussia. When the dust settled, Germany would emerge as the undisputed superpower of Europe.

The Schlieffen plan almost worked. The operative word here is 'almost.' After overwhelming Belgium and marching into France, the German offensive stalled a few miles outside Paris.

Then, contrary to German expectations, Great Britain entered the war on the side of France and Russia. To top it off, Russia mobilized its forces faster than the Germans had anticipated. At this juncture, the Germans made a fateful decision to counterattack on the Eastern front, gambling on a quick victory followed by a Russian withdrawal.

In August 1914, Germany devastated the Russian army at Tannenberg, but to the Germans' amazement, Russia did not withdraw from the war. Instead, the Czar ordered wave after wave of fresh units to the Eastern front. Although Russia's army was composed mainly of malnourished, poorly armed peasants, their sheer numbers prevented Germany from hurling its entire force at the British and French, who were solidly entrenched on the Western front by then.

The Sinking of the Lusitania

At the start of the war, the United States was officially neutral but soon began shipping arms and supplies to the British. In response, the Germans set up an Atlantic naval blockade and engaged in unrestricted submarine warfare. Then, in 1915, a German U-boat sank the luxury liner Lusitania, killing nearly 1,200 people, 128 of them Americans.

Many Americans clamored for President Wilson to enter the war against Germany right then and there. Instead, Wilson bought time by warning Germany multiple times to cease all submarine attacks on vessels flying the U.S. flag.

The Zimmermann Telegram

Strictly speaking, this incident was a diplomatic blunder; nonetheless, it proved to be the proverbial last straw in provoking the U.S. to declare war against Germany. In January 1917, British agents intercepted an encrypted telegram from Arthur Zimmermann, the Foreign Secretary of the German Empire, to von Eckhardt, the German minister to Mexico. The message stated that if Mexico attacked the U.S., Germany would one day help Mexico regain the territory it had ceded to the U.S. under the treaty of Guadaloupe-Hidalgo in 1847.

Predictably, much of the U.S. public was outraged when the press broke the story of Germany's plot to help Mexico conquer U.S. territory. In hindsight, this fiasco reflects Germany's growing desperation to avoid direct conflict with the U.S. Less than a month after the telegram was intercepted, Germany resumed unrestricted submarine warfare, making a showdown with the U.S. inevitable.

By the time Germany faced U.S. troops; however, it had already suffered over 3.5 million casualties, amounting to nearly one-third of its entire army. Although Russia bowed out of the war shortly after the U.S. jumped in, Germany was no match for the combined strength of the American, British, and French forces. By August 1918, German forces fell back from the Western Front; a few weeks later, German troops retreated from Romania and the Balkans. An outraged German public pressured Kaiser Wilhelm II to abdicate, and on November 11, 1918, a provisional German government signed the armistice at Versailles, ending WWI.

Essay 117: From Hunter-Gatherers to Farmers

How did humans learn to farm? The question is deceptively simple, but solving this riddle took years of effort by scholars from backgrounds as diverse as anthropology, archaeology, botany, genetics, and physics. As we shall see, the questions of when, where, and how agriculture began had profound implications for the course of human history.

When

The general consensus is that all humans on Earth lived as hunter/gatherers until the end of the last Ice Age, approximately 12,000 years ago. At that point, conditions in certain parts of the world favored plant domestication over hunting/gathering. Grasslands expanded in the Fertile Crescent (the arc of land running from Israel through Turkey east to Iran) and China. Contemporaneously, human hunters in these regions (and elsewhere) depleted large animal species.

Ultimately, long-distance trade networks, which allowed for the spread of domesticated plants and animals, linked Asia, North Africa, and Europe and arose at later times in other parts of the world. By radiocarbon dating, scientists think the transition to agriculture took place around 8500 B.C.E. in the Fertile Crescent, a few centuries later in China, and at later times in most of the rest of the world. Today, a few bands of hunter/gatherers survive in areas unsuitable for agriculture, such as the Arctic, the Kalahari Desert, and the lowland swamps of New Guinea.

Where

In the Old World, the Fertile Crescent and China are thought to be the earliest sites of agriculture. A few thousand years later, agriculture arose independently in New Guinea and probably in West Africa as well. In the New World, agriculture arose independently in Mexico, the Eastern United States, and the Andes mountains of South America.

Agriculture spread rapidly from its core areas in the Old World (aside from the New Guinea highlands) but far more slowly in the New World. According to the biologist Jared Diamond, the fundamental reason for this discrepancy is the geographical orientation of the continents.

In his epic work Guns, Germs, and Steel, Diamond notes that the major axis of Europe and Asia runs east to west, whereas the axes of Africa and the Americas run north to south. Consequently, while similar crops can grow from the Mediterranean to China, the same is not true for most crops in the Americas (and, to some extent, sub-Saharan Africa). Variations in day length, average temperature, and seasonal rainfall patterns conspired to limit most New World crops, other than corn and beans, to their sites of origin.

How

The wild ancestors of wheat, barley, chickpeas, peas, and flax are native to the Fertile Crescent, while those of soybeans and rice are native to China. All of these plants proved to be comparatively easy to domesticate as annual crops. In contrast, corn, the leading crop in New World agriculture, arose from teosinte, a plant consisting of a tiny cob with a few inedible seed kernels. While Old World crops evolved into modern forms in a few generations, the transition from teosinte to corn took centuries, possibly even millennia.

The Consequences

The first societies to adopt large-scale agriculture (often along with animal domestication) expanded in population and became politically, economically, and militarily dominant in their respective regions of the world. It is no accident that most empires in history arose along the east-west axis of the Mediterranean, the Middle East, India, and East Asia.

In the Americas, the Maya city-states arose by 200 B.C.E. By the 15th century CE, two large states developed in the form of the Aztec and Inca empires. Nevertheless, the later adoption of agriculture delayed the rise of these New World states by 3,000 years compared to their Old World counterparts, Egypt and Sumer. By 1492, Europeans had developed ocean-going ships; this advance made Spanish and Portuguese voyages to the New World all but inevitable. Over the next century, England, France, and Holland would join in the conquest of the Americas.

Indeed, the when, where, and how of agriculture's origins have profoundly shaped the human experience for over 10,000 years.

Essay 118: International Security in an Era of Globalization

A political science professor once commented, 'We can only understand political power and international relations with a knowledge of resource distribution and technology.'

This statement provides a framework for understanding the formation of nation-states culminating in the rise of global superpowers in the 20th century. Building upon this foundation, it is critical to consider other factors that permeate modern international relations, particularly alliances (and rivalries) rooted in the political, economic, cultural, and religious histories of the nations involved. This essay will explore the role each of these institutions has played and continues to play in shaping the balance of global power.

Understanding how we got here: A basic knowledge of geography is fundamental to understanding the broad patterns of human history and, ultimately, international relations today.

All states and empires in recorded history arose in the past 5,500 years in regions of the world with the most advanced systems of food production and largest human populations. As we shall see, it is no coincidence these areas include the Mediterranean basin, the Middle East, India, China, and Southeast Asia. Empires also arose in parts of Africa south of the Sahara, including Ethiopia, the Niger River of West Africa, and regions stretching from Kenya to the Cape of Good Hope. In the Americas, the Aztec and Inca empires arose by the 15th century AD, some 4,500 years after the rise of city-states in Sumer (Mesopotamia) and the Old Kingdom in Egypt.

This delay in the rise of large-scale food production in the New World compared to the Old World had profound consequences for international relations after the year 1500, especially with regard to the rapid conquest and colonization of most of the Americas by Europeans. In the case of sub-Saharan Africa, European colonization occurred three and a half centuries later than in the Americas for a few key reasons: food production, metallurgy, and advanced political states had a longer history in Africa compared to North and South America. Moreover, tropical diseases, especially malaria, decimated European colonists in Africa until quinine came into widespread use in Western medicine during the 19th century.

Jared Diamond's 1997 masterpiece Guns, Germs, and Steel is based on this premise: The greater variety and abundance of plants and animals suitable for domestication in Europe and Asia compared to the rest of the world coupled with the East/West geographical axis of Eurasia vs. the North/South axis of Africa and the Americas has had the most profound impact on human history in terms of relative population sizes, development of technology and weapons, and subsequent political power.

Farming allowed human populations to grow large in certain parts of the world – in particular, the Mediterranean, the Middle East, India, and China, all of which became linked by trade networks over 2,000 years ago. Metallurgy developed early in the Middle East and China, then gradually spread throughout most of the Old World, with the exception of New Guinea, Australia, and Polynesia. In addition to agriculture, animal domestication, and metalworking, other inventions, both ancient and more recent, spread quickly along the East-West axis of the Old World. Examples include the wheel, writing, ceramic pottery, maritime technology, paper, compass, and gunpowder.

In contrast to the swift spread of animals, plants, and technology throughout mainland Eurasia, other parts of the world lagged behind due to their remote locations, limited animal and plant resources, and/or later dates of human settlement.

Polynesia / New Guinea – Polynesians are believed to have migrated from Taiwan and Malaysia, reaching all habitable islands in the Pacific by around the year 1000 AD. While they brought pigs, dogs, chickens, and tropical Southeast Asian crops with them, Polynesian societies lacked writing and metal tools. Agriculture developed independently in New Guinea, but except for some elements of Polynesian food production (chickens and pigs), contact between this large island and the rest of the world was extremely limited.

Within New Guinea itself, mountain ranges and dense rainforests made long-distance travel difficult and political unity impossible. The population lived in small villages or as nomadic groups speaking hundreds of unrelated languages. Of the 6,000 surviving languages in the world today, over 800 are spoken only in New Guinea. Until the 1930s, the tribes of the central New Guinea highlands, a remote region surrounded by high mountains and impenetrable jungle, remained completely unknown to the outside world.

Aboriginal Australia – Australia is the most remote continent and the least hospitable for food production. Native Australians never developed agriculture in the traditional sense; some groups intentionally burned the underbrush, promoting edible fern growth. Other Aboriginal bands built elaborate canals along the Murray-Darling River that served as eel fisheries.

Over 75% of Australia is desert. The most suitable areas for agriculture are the southeast and southwest corners of the continent. However, the plants that grow best there are temperate zone crops (wheat, oats, apples, grapes), not tropical ones. Temperate zone crops only arrived after the British landed in 1788.

Within a century of the British arrival, most Aborigines died from epidemic diseases or European guns. The survivors were treated as second-class citizens or driven into desert regions of Australia unsuitable for farming or sheep herding.

Sub-Saharan Africa – The Sahara desert is a major barrier to human migration; by medieval times, Arab caravans became the major trade link between West Africa and the rest of Eurasia. By the 14th century, Mali, with its capital at Timbuktu, dominated the region through the gold-salt trade. States in West Africa adopted horses but never made it through the tsetse fly zone of equatorial Africa. Metalworking and ceramic pottery reached South Africa by the year 1 AD, thousands of years after Egypt and other parts of North Africa had mastered these skills. Farming, cattle, and metal weapons allowed Bantu-speaking people from tropical West Africa to spread throughout southern Africa by the year 1000 AD.

For the reasons mentioned above (malaria/endemic tropical diseases), Europeans conquered sub-Saharan Africa only after 1850. Except for South Africa (which has a temperate climate and lacks tropical diseases), large numbers of Europeans never settled in Africa during the colonial era. Less than a century later, most of sub-Saharan Africa regained its political independence. Internal political disputes, chronic civil war, deforestation, explosive population growth, famine, drought, AIDS, and other destabilizing socioeconomic factors represent Africa's biggest challenges in the post-colonial era.

Mesoamerica, Eastern U.S., and the Andes – Farming developed in each region, but its spread was extremely slow. Mexican crops took centuries to reach the Eastern U.S. and never reached South America until after the arrival of Europeans. Similarly, potatoes remained confined to the Peruvian highlands until after the Spanish conquest of 1532. The underlying reasons come down to geography and climate. Potatoes do not grow well in lowland tropical regions; Mexico and Peru are separated by 1,200 miles of terrain inhospitable to highland crops. The dense rainforest blanketing Central America and the Isthmus of Panama blocked overland trade between Mexico and South America. Trade routes connecting North American centers of food production were tenuous at best.

With no domesticated animals that could carry a rider, long-distance travel and routine communication never developed over large areas of the New World.

The two largest food-producing regions of the Americas gave rise to the Aztec and Inca empires; however, both empires relied on stone tools for farming and human runners to carry information. After the arrival of Europeans, both empires fell quickly to tiny armies of Spanish conquistadores.

That represents a capsule summary of world history up to the modern era. In the afterword of Guns Germs and Steel, Diamond emphasizes the profound legacy of food production and geography on the course of human history:

"The nations rising to new power are still ones that were incorporated thousands of years ago into the old centers of dominance based on food production, or that have been repopulated by peoples from those centers…The world's two earliest centers of food production, the Fertile Crescent and China, still dominate the modern world, either through their immediate successor states (modern China) or through states situated in neighboring regions influenced early by those two centers (Japan, Korea, Malaysia, and Europe), or through states repopulated or ruled by their overseas emigrants (the United States, Australia, Brazil)."

Modern Political Alliances: WWII, the Cold War, and the balance of power in the 21st century.

Continental axes, large populations, surplus food production, metal working, and maritime technology set the stage for the expansion of overseas European colonies after 1500. These factors ultimately paved the way for the rise of regional and global powers in the form of Great Britain, France, Germany, the U.S., Russia, India, China, and Japan. The empires of Western Europe peaked in the early 20th century and then declined after World War II. Today, it is almost impossible to understand the political landscape of the modern world without a solid grasp of World War II and its aftermath.

During WWII, the Allied powers, namely the U.S., Great Britain, and the Soviet Union, joined forces to defeat Nazi Germany and the Empire of Japan.

As the war raged, the U.S. assembled the world's most renowned physicists to develop an atomic bomb in a secret program known as the Manhattan Project. In August 1945, President Truman ordered atomic bombs to be dropped on Hiroshima and Nagasaki, forcing Japan to surrender unconditionally and ending WWII abruptly.

Within a year of the war's end, the alliance splintered, and most countries gravitated either toward the Western powers - the U.S., U.K., and eventually the North Atlantic Treaty Organization (NATO) - or else became Soviet allies or satellites, in the case of most of Eastern Europe. A few countries, most notably India, designated themselves 'non-aligned' or neutral nations during the Cold War era. Meanwhile, U.S. foreign policy crystallized around the doctrine of containment – to fight Soviet domination and the spread of communism by any means short of all-out war against the USSR itself. Paradoxically, the existence of nuclear weapons and the threat of a global cataclysm stopped the Cold War from heating up.

Although some scientists and intellectuals cherished the hope that no other country besides the U.S. would seek to develop nuclear weapons, the opposite scenario occurred. By 1949, the USSR had become a nuclear power. Britain followed in 1952 and France by 1960. In 1964, China detonated its first atomic bomb. A decade later, India and Pakistan developed nuclear weapons. Today, it is uncertain how many nations possess nuclear weapons. The consensus is that Israel developed nuclear weapons by the 1960s, although the Israeli government has never officially acknowledged such a program. Some former Soviet Republics probably possessed nuclear warheads deployed on their soil prior to the breakup of the USSR in 1991.

North Korea most likely possesses nuclear warheads; however, the number and range of these weapons remain an open question.

Although the sharp political divisions of the Cold War era have blurred or faded into irrelevance, key alliances from that period of time remain intact. The U.S. camp still includes Western Europe, all NATO states (e.g., Turkey), much of Latin America, Japan, South Korea, Indonesia, Australia, New Zealand, and several states in the Middle East, especially Israel, Jordan, Kuwait, Qatar, Saudi Arabia, and the United Arab Emirates.

Iraq and Afghanistan represent special cases in the sense of having governments installed under U.S. military occupation and still supported by a limited number of U.S. and allied troops.

The pro-Russia camp (largely an extension of the former Soviet bloc) includes a few former Soviet republics, the People's Republic of China (which has arguably become a superpower in its own right), as well as other surviving communist and socialist regimes – North Korea, Cuba, and Venezuela being the most ardent supporters. Certain other regimes have chosen to ally themselves with Moscow, namely Iran, and Syria, but largely as a reactionary measure to oppose U.S. interests. Traditionally, non-aligned nations like India have gravitated toward the U.S. camp, while China has increased its economic leverage in regions like sub-Saharan Africa, much of which remains mired in political turmoil.

Hence, the international political order of the 21st century still bears a great deal of resemblance to the Cold War era half a century ago. Beyond a tacit acceptance of the American, Russian, and Chinese spheres of influence, there are few, if any, rules that apply in the modern international realm. Given its abysmal track record at resolving transnational conflicts, it is no exaggeration to say the United Nations rapidly became an irrelevant debate club after the end of the Korean War in 1953. Moreover, the nebulous treaties that constitute international law are basically unenforceable without the cooperation of the U.S., Russia, and China.

The major paradigm shift is that economic ties usually trump political philosophies in the modern world. Therefore, it is vital to consider the balance of global economic power in our analysis of international relations.

Economics – it still boils down to Food and Energy

The availability of food and energy resources and the ability to harness them form the foundation of the global economic order. This may initially sound like a radical idea, but its basic truth remains self-evident, to paraphrase Thomas Jefferson. Although many countries around the world are food self-sufficient, relatively few are exporters of staple foods like wheat, corn, or rice. Many countries in Asia, Africa, Latin America, and the Caribbean depend on grain imports from the U.S., Canada, Australia, and Russia.

There is a less clear-cut relationship between global economic power and rice exports, led by Thailand, Vietnam, Pakistan, and India. In the past three years, China has become the second leading importer of rice after Nigeria; however, China still leads the world in rice production, growing an estimated 143 million tons in 2013. Aside from this trend, other economic powerhouses in Asia, like Taiwan, Japan, and South Korea, grow sufficient rice to meet their needs without becoming major importers or exporters of this staple grain.

Beyond staple foods, certain commodities like bananas, coffee, tea, and cocoa beans bolster the economies of several countries, including Brazil, Colombia, Sri Lanka, Indonesia, and the Ivory Coast. As much as people enjoy chocolate, coffee, or tea, there is still a sharp distinction between these products and staple foods. If all trade in, say, coffee and cocoa beans stopped entirely, many people in the U.S. and Europe would no doubt be upset but could still obtain caffeine from colas or tablets sold in pharmacies. In contrast, if global trade in wheat and rice were suspended, millions of people around the world would face starvation in a matter of months.

The other major pillar of global power is energy, specifically energy self-sufficiency, to a greater extent than oil exports. This may seem counterintuitive, but consider these trends. The U.S. imports less oil today than it did in the 1980s, even though the U.S. population has grown by over 30 million.

Russia remains a net exporter of oil and natural gas, with which it will continue to dominate Eastern Europe and Central Asia economically. China and India will continue to import most of their oil but are steadily transitioning to coal and natural gas as well.

According to a 2010 article in National Geographic, world oil production may have peaked in 2006. Oil will continue to dwindle in the coming years as existing fields are depleted. At the same time, new reserves tend to be located in remote or prohibitively expensive areas to drill (e.g., under the ocean floor).

Simply put, the days of the OPEC cartel are numbered as the global economy shifts toward coal, liquefied natural gas, and nuclear energy as the main sources of power for the next several centuries. Biofuels, solar power, hydroelectric, wind, and geothermal energy will also become increasingly vital to the world economy because, unlike fossil fuels, they are renewable sources of energy.

Social

Differences in language, culture, and religion have traditionally been sources of conflict in the realm of international relations. Until the last few thousand years, barriers in the form of unrelated languages and obstacles to travel kept much of the world in a state similar to New Guinea of the 1930s; people seldom traveled more than ten miles from their place of birth during their entire lives. Increased communication in the form of trade networks arose first in the Middle East and China, later in Europe and Sub-Saharan Africa, and later still in North and South America.

Although language and cultural barriers exist to a certain extent today, they come into play mostly at the level of local and regional conflicts (especially religious differences) and are not the ultimate determinants of the global balance of power.

Religion continues to play an important role in many people's personal lives and, to some extent, shapes national identity (Catholicism in Ireland and Islam in Saudi Arabia). In and of itself, however, religious identity has never determined power on the international level to the same extent as military strength and economic prowess.

For example, most European nations have had a Christian majority for over 1,000 years. Similarly, most nations in the Middle East and Central Asia have had Muslim majorities, and Buddhism continues to be the predominant religion in East Asia. During this entire period of time, wars within and between those regions of the world have not been decided based on the combatants' religious identity but rather by the side with the most well-equipped army, the best-organized supply lines, and access to advanced weaponry.

The Internet – the great equalizer?

On one level, the answer is yes, in the sense that the internet has connected people like never before. This revolution in communications is still in its early stages; a mere 20 years ago, websites like Google and Facebook did not exist. Whether or not the cyber age will redistribute the global balance of power remains to be seen. As of now, the internet has put vast amounts of information at the fingertips of anyone with access to a computer. However, this access remains mostly limited to people in affluent countries with the political and economic strength that made the infrastructure underlying the Internet possible in the first place.

In a sense, the phenomenon of instant communication via instant messaging, email, or Skype is merely an extension of the telephone, invented in 1876. Online information archives are supplanting physical collections once housed in libraries; nevertheless, books, magazines, and newspapers remain popular in both traditional and electronic formats. Not coincidentally, the nations with the strongest traditions of mass literacy and freedom of information still hold a decisive advantage in this realm. It seems like the cliché, "the more things change, the more they stay the same," is really a profound statement of truth.

Essay 119: The Legacy of the Thirty Years' War

The historian Thomas Munck once remarked that the Thirty Years War achieved little 'beyond making permanent a political compromise just about reached in 1555'. Although this statement accurately describes the political situation in the German States and the Hapsburg Empire after 1648, it fails to consider the wider ramifications of the Thirty Years War for the rest of Europe. This essay will compare and contrast the balance of power in Europe in 1555 and a century later in 1648, and discuss why Munck's analysis misses the mark when describing the legacy of the Thirty Years War.

Understanding the 1555 Peace of Augsburg requires some background as well. Nearly four decades prior to this event, Martin Luther launched the Protestant Reformation in 1517 by publishing his work The 95 Theses. This scathing indictment of the corruption entrenched in the Catholic Church brought long-standing tensions to a head throughout Northern Europe. In the span of one lifetime, much of Germany, England, Scotland, and Scandinavia broke away from the Roman Catholic Church. Switzerland became an enclave of Calvinism. Although the Vatican launched a Counter-Reformation to combat Protestant sectarianism, the hegemony of the Catholic Church over Western Europe was permanently broken.

Given that 16th-century Europe was accustomed to intermittent warfare, it is not surprising that violence erupted between Catholics and Protestants for much of the following century. In England, Henry VIII dissolved monasteries and confiscated lands held by the Catholic Church. In France, Protestant Huguenots were persecuted and then massacred by the thousands in Paris in 1572. In the Netherlands, a religious truce held as the Dutch (both Catholic and Protestant) broke away from Spain and established an independent state in 1609.

Germany also saw decades of conflict, but unlike the aforementioned countries, it had been disunited since the end of Charlemagne's reign in 814. A nominal ruler in the guise of the Holy Roman Emperor served as the figurehead monarch of Germany; however, real political power was divided between the Habsburg Empire in Austria and hundreds of feuding principalities in Germany. Augsburg's peace (or rather truce) served as a stop-gap measure to end the fighting among German states by adopting a few key measures: First, German princes would dictate the official religion of their principalities. Second, Catholics or Protestant landowners who emigrated from one territory to another to avoid converting to the ruler's religion would be compensated (ideally) for the value of the relinquished property. Finally, all German princes were expected to nominally recognize the authority of the Holy Roman Emperor (a compromise designed to maintain the status quo).

The major accomplishment of the Peace of Augsburg was that German princes maintained an uneasy truce for the next 60 years under the figurehead authority of the Holy Roman Empire.

In the larger context of European politics, however, it is critical to note what did not take place before 1555 - invasions by the states surrounding Germany in support of Catholic or Protestant factions in the conflict. The Catholic-Protestant wars in 16th century Germany remained an internal power struggle for reasons to be discussed next. The key difference in the Thirty Years War was that Spain, France, Denmark, Sweden, Poland, and the Papal States were all too willing to plunge into the renewed violence engulfing Germany.

While historians may debate the reasons for the lack of external intervention in Germany during the early decades of the Reformation, several straightforward explanations should suffice. First, the age of exploration was in full swing.

Spain and Portugal, followed by France, England, Denmark, and Sweden, vied with one another to gain colonies in the New World. This preoccupation with conquest and capturing a share of the wealth flowing into Spain pushed any concern about German religious conflict to the back burner, so to speak.

Second, aside from the stability of the Spanish monarchy and the Tudor Dynasty in England, the rest of Europe was in almost as much turmoil as Germany during the early decades of the Reformation. In France, a weak monarchy reigned in Paris as the nobility contended for power. In Italy, city-states fought against French and Spanish encroachment, dominance over Mediterranean trade routes, and control over the obsolescent Silk Road. In Eastern Europe, the Hapsburgs could barely contain Ottoman expansion as the Turks reached the gates of Vienna in 1529. Meanwhile, Poland and Russia fought intermittently to expand their spheres of influence in Ukraine. In short, few 16th-century rulers were interested in meddling in Germany's internal power struggle.

The European political and economic landscapes changed significantly by the early 17th century. The major European powers had drifted into Catholic and Protestant alliances; this set the stage for a continent-wide power struggle. In 1609, the Dutch won independence from Spain (with the exception of the Spanish Netherlands). In a few years' time, the Dutch East and West India companies established a global maritime trade network unparalleled in the rest of Europe. Soon, the Dutch would consolidate their territories in the Caribbean and Indonesia with a powerful sailing fleet and reliance on mercenaries. Although a religious truce was held in the tolerant environment of the Netherlands, as a matter of policy (and to get even with Spain), the Dutch supported the Protestant side in the Thirty Years' War.

In England, the monarchy remained pro-Catholic through the reign of Charles I; however, Protestant forces gained the upper hand by the 1640s and triumphed in the English Civil War. England grew stronger in European affairs after the Spanish Armada's defeat in 1588. At the same time, Spain, a staunch defender of Catholicism, was declining in military prowess. Part of this was due to its defeat in 1588. Other factors included the rise in piracy in the New World as French privateers prowled the western coast of Hispaniola while English pirates wrested Jamaica, Barbados, and other Caribbean islands from Spanish control. In South America, silver output declined as mines were depleted. Except for the Philippines, Spain had no other significant colonies outside of the Western Hemisphere. Meanwhile, the Dutch gained territory in Indonesia while the British, French, and Scandinavians raided Spanish treasure fleets in the Atlantic.

France is another important power to consider in the context of the Thirty Years' War. Although the French monarchy remained Catholic (and persecuted Protestants inside the borders of France), France was alarmed at the prospect of encirclement by a Hapsburg dynasty ruling in both Spain and Austria (which, by extension, meant the Holy Roman Empire). France reacted by supporting the Protestant side during the Thirty Years' War.

Denmark and Sweden vied for control of the Baltic in the 16th century but, by the 1600s, became staunch defenders of the Protestant cause. Although neither country acquired as many overseas colonies as Britain, France, or the Netherlands, their dominance of Baltic trade routes and military prowess proved critical during the second half of the Thirty Years War. Geographically, Sweden was positioned to send troops to fight in Germany, whereas the Hapsburgs and Holy Roman Empire never managed to construct a naval fleet that posed a serious threat to Sweden. As such, Swedish forces under Gustavus Adolphus waged an offensive war on Germany rather than vice versa.

In contrast to the Protestant camp, Catholic forces held some key advantages at the start of the war (although these declined in importance as the war ground on). First, Spain, Italy, Austria, and Poland border Germany or can maintain direct land routes.

France, Denmark, and the Netherlands also shared land borders, which guaranteed that Germany would be ravaged by the war in contrast to the surrounding nations, which saw little or no fighting on their soil.

Second, the Catholic alliance was arguably more united in its loyalty to the Papacy and dedication to the Counter-Reformation than the Protestant side was in its support of various German states. This may account for several Catholic victories prior to 1630. As the war progressed, however, Denmark and Sweden regained military momentum and recaptured most of northern Germany.

As for the Thirty Years War outcome, Germany suffered vastly more during this time than it had in the years leading up to 1555. In the 16th century, wars had been shorter and more sporadic - more like skirmishes than prolonged invasions.

Armies fought each other fiercely but avoided large-scale massacres of civilians (at least compared to the medieval era). The Thirty Years' War was a different story. Foreign armies roamed much of Germany, pillaging villages and fields. Rather than simply disband and return home after battles, soldiers on both sides turned the German countryside into a scene of non-stop plunder.

Beyond the borders of Germany, the balance of power in Europe shifted in favor of the Protestant alliance. France became the strongest military power in Western Europe, a position it enjoyed through the Napoleonic Wars of the 19th century. England continued to colonize the New World and eventually embarked upon the conquest of India. To avoid further religious conflict in politics, Parliament elevated the Church of England to official status and later outlawed the possibility of a future Catholic monarch after the death of Queen Anne in 1714. Dutch economic power peaked in the 17th century, as did Danish and Swedish military power. All in all, Northern Europe emerged stronger economically and militarily after 1648 compared to the Catholic states of Southern and Central Europe.

The historical trajectories of Spain, Italy, Poland, and the Hapsburg Empire were drastically different after 1648. Spain continued to decline as a military and economic power in Western Europe and the Americas. Italy remained divided among quarreling city-states until the late 19th century. Just as the Thirty Years War ended, Poland was devastated by a Cossack uprising led by Bogdan Chimielnicki. For the next century and a half, Poland suffered a series of invasions by Tatars, Swedes, Austrians, and Russians. By 1792, Poland lost its independence completely as it was partitioned between Prussia, Austria-Hungary, and Czarist Russia. The Hapsburg empire lost its position of prominence as Germany gradually recovered from the Thirty Years War. By 1866, any hegemony the Hapsburgs enjoyed over Germany came to an end with Austria's defeat in the Seven Weeks War. Finally, the Holy Roman Empire was already in steep decline until 1806, when Napoleon officially abolished it.

Conclusions:

The results of the Thirty Years War were devastating but predictable. Between 25 to 40 percent of Germany's civilian population perished during the course of the war. The main causes (beyond pitched battles) were starvation and epidemics. Germany's population did not recover to its 1618 level for over a century. On the economic front, Germany was ruined and had no chance of gaining overseas colonies as the Age of Exploration drew to a close. Politically, any possibility of German unity was pushed centuries into the future. In the interim, Germany remained divided along the religious borders laid down at Augsburg in 1555. While the regional power of Prussia grew in the 18th century, political unification required another century and the leadership of Otto Von Bismarck. Chronic disunity, distrust, and division gripped Germany from 1648 until the Franco-Prussian War of 1870. This was the true legacy of the Thirty Years' War.

Essay 120: The Seven Wonders of the Ancient World

The Seven Wonders of the Ancient World are mentioned in Herodotus's writings and later by antiquity historians. Most lists include 1) The Great Pyramid of Cheops (or Khufu), 2) The statue of Zeus at Olympia, 3) The Colossus of Rhodes, 4) The Temple of Artemis at Ephesus, 5) The Mausoleum at Halicarnassus, 6) The Hanging Gardens of Babylon and 7) The Pharos (or lighthouse) of Alexandria. In some lists, the Walls of Babylon take the place of the Pharos of Alexandria. Of these seven ancient wonders, the only one still in existence today is the Great Pyramid in Egypt, completed circa 2800 BCE.

This essay will track the fates of the other ancient wonders, with emphasis on the Colossus of Rhodes and the Pharos of Alexandria. Except for the Great Pyramid, these two wonders existed for the shortest and longest periods, respectively.

The Hanging Gardens of Babylon were built on the orders of King Nebuchadnezzar, who ruled Babylonia during the 6th century BCE. The story goes that the king's favorite wife came from a mountainous region bordering Mesopotamia and was homesick, so Nebuchadnezzar ordered trees and flowers planted on the terraces of one or more of the ziggurats in Babylon. It is unclear what became of the gardens between the capture of Babylon by the Persians in 539 BCE and Alexander the Great's conquest of the city two centuries later. By the first century BCE, Babylon itself had become little more than a pile of ruins following a series of droughts and earthquakes.

The Statue of Zeus at Olympia was carved by Phidias around 432 BCE, near the site of the ancient Greek Olympic Games. These games continued all the way until the year 392 CE when the Roman emperor Theodosius II prohibited pagan festivals throughout the Eastern Roman Empire. No one is sure what happened to the statue of Zeus. Because the statue contained a great deal of gold and ivory, it was no doubt plundered; whether by Gothic raiders or the Byzantines themselves remains uncertain. The shrine containing the statue of Zeus is believed to have burned down in the year 425 CE.

The Temple of Artemis was built in Ephesus (now in modern-day Turkey) in 550 BCE. It stood until 262 BCE when Gothic invaders destroyed it.

A few decades before the Temple of Artemis was built, construction began on the Mausoleum at Halicarnassus, built by Queen Artemisia as a tomb for her husband Mausolus following his death in 353 BCE. The mausoleum survived until the 14th century BCE, when it was destroyed by a series of earthquakes. By the year 1402, only the foundation of the mausoleum remained. The Knights of St. John are thought to have used some of its stones to build their fortress.

The Colossus of Rhodes was constructed around the year 292 BCE by the sculptor Chares to commemorate the siege of Rhodes by Demetrius, the son of Antigonus, a general under Alexander the Great.

Rhodes withstood the year-long siege and emerged triumphant after Demetrius's forces withdrew following the arrival of Ptolemy II's forces from Egypt. The retreating army left an immense siege tower behind, which became the scaffold of the Colossus.

Twelve years later, a towering statue covered in bronze stood where the siege tower had once been. Depicting the ancient Greek sun god Helios, the Colossus stood an estimated 110 feet high.

Contrary to popular myth and several Renaissance paintings, most archaeologists think the Colossus was built adjacent to the harbor at Rhodes; it did not straddle the harbor's entrance.

The statue was destroyed in an earthquake in 226 BCE, a mere 66 years after it was completed. The people of Rhodes supposedly believed the earthquake was an omen, that the Gods were displeased, and the Colossus was never rebuilt. Nine centuries later, Arab salvage crews recovered much of its metal from the bottom of the harbor.

The Pharos of Alexandria was built starting in 290 BCE during the reign of Ptolemy II, the second king in a dynasty that ruled Egypt following the breakup of Alexander's Empire. The designer of the lighthouse is believed to have been Sostratus of Knidos. This lighthouse was built on a small island called the Pharos and, when completed, towered an estimated 450 feet above the harbor at Alexandria. A few sources claim the Pharos stood 600 feet high. If they are correct, it would mean the Pharos stood even taller than the Great Pyramid at Giza, which held the record for the world's tallest man-made structure until the Eiffel Tower was built in 1889.

Although no one is certain about the nature of the fuel burned at the top of this gigantic lighthouse, some ancient sources claim its light was visible 50 miles out to sea. According to one theory, polished sheets of metal were used like giant mirrors to amplify the light. Historians think the Pharos survived relatively intact until the year 1303 when its top 70 feet came tumbling down during an earthquake. By 1375, much of the remaining structure crumbled. The Ottomans used Many of the stones a century later to build a fortress where the lighthouse had once stood. Some of the stones from the Pharos remained submerged in Alexandria's harbor until their rediscovery in 1994.

Essay 121: Who Won the Cold War?

The Cold War started less than a year after the end of World War II. The Soviet Union, still under the rule of Josef Stalin, occupied Eastern Europe and installed so called satellite governments in Poland, Czechoslovakia, Hungary, Bulgaria, Romania and Albania. Besides controlling its satellites, the USSR also gave financial and military support to communist insurgencies in Greece, Turkey, China, Korea, and later in Southeast Asia. In a famous speech delivered in 1946, former British Prime Minister Winston Churchill declared that "From Stettin in the Baltic to Trieste in the Adriatic, an Iron Curtain has descended across the Continent." (Rosenberg, 2014). By the 1950's the Iron Curtain divided much of the world along geographic and ideological lines.

Western Europe and Canada, other members of the British Commonwealth, and Japan remained solid allies of the United States. Latin American governments uniformly sided with the U.S. until the Cuban revolution of 1959. Countries with historic ties to the U.S. like the Philippines and Liberia remained steadfast allies as well. Finally, certain countries or nationalities that had suffered oppression under Czarist Russia or who staunchly opposed the Soviet Union's militant atheism and suppression of religious freedom gravitated toward the U.S. camp. These allies included Israel, Turkey, Iran, Saudi Arabia, and Indonesia.

As the U.S. consolidated its global alliance, the Soviets made every effort to bring neighboring countries into the Kremlin's sphere of influence. In the early years of the Cold War, these efforts yielded a few tangible results. First, in 1949 communist forces under Mao Zedong emerged victorious in China's civil war. A year later, the People's Republic of China sent troops into North Korea, initially to help communist troops overrun South Korea, and eventually to rescue the North Korean government from a U.S. led counter invasion. Second, in 1955, the USSR's Eastern European satellites formed the Warsaw Pact, an organization meant to counterbalance the North Atlantic Treaty Organization (NATO). Although the Warsaw Pact voted as a unanimous bloc in the United Nations, it represented a symbol of communist unity more so than a military threat (History.com, 2014).

Under President Truman, the U.S. responded to communist expansion in general and Soviet backed militarism in particular with a policy called containment – supplying any opposition movements to communism with money and weapons. Diplomats, military advisers, mercenaries, and CIA agents participated in the Cold War at different times and places. As a last resort, the U.S. sent American troops to fight Communist forces directly.

The two major U.S. military interventions during the Cold War were Korea, which lasted from 1950 to 1953, and Vietnam, where direct U.S. military intervention lasted from 1965 to 1973. Although each conflict affected the balance of global military power in the short term, neither one determined the ultimate outcome of the Cold War. Instead, the Soviet Union's disastrous decision to invade Afghanistan in 1979 combined with a moribund economy suffocating under Communist mismanagement triggered the collapse of the USSR and the end of the Cold War era. The remainder of this essay will analyze the key events of the Cold War in the hopes of providing a meaningful answer to the title question.

In 1948, Berlin became the first flashpoint of the Cold War. The Russians blockaded West Berlin (the allied sector of the city). In response, Truman ordered Operation Victuals, better known as the Berlin Airlift, to continuously supply the city with food, medicine, and fuel for over a year. The crisis ended when the Soviets lifted the siege (Truman Library, 1988). In 1961, however, the Soviets erected the Berlin Wall to prevent East Germans from defecting to West Berlin. The Berlin wall stood until 1989; its collapse would mark the symbolic end of the Cold War and the beginning of the end of Russian hegemony over Eastern Europe.

In 1950, Chinese and North Korean forces, supplied by the Russians, invaded South Korea. This sparked the Korean War, which cost the lives of 50,000 Americans and other U.N. troops.

Although the communist regime in North Korea was not toppled, South Korea was saved, and Truman's doctrine of containment was vindicated for the time being. From the perspective of the USSR, the Korean War was a draw. The Soviets lost military equipment, but it was China that sacrificed 500,000 troops to prop up a communist regime in North Korea (U.S. History, 2014). As the Cold War wore on, military cooperation and ultimately diplomatic relations between the Soviets and Chinese grew strained.

Unlike Truman, President Eisenhower favored covert CIA actions (e.g. installing or bolstering pro-American dictatorships in Latin America and Iran) along with economic support to check the spread of communism. John F. Kennedy more or less continued these policies. He resolved the Cuban Missile Crisis in 1962 by striking a secret deal with Khrushchev to remove U.S. missiles from Turkey in return for the Soviets removing missiles from Cuba. Many historians view the Cuban Missile Crisis as the climax of the Cold War in the sense that the U.S. and U.S.S.R. never came as close to direct military confrontation afterwards. Other scholars note the paradox at the heart of the Cold War – both sides wanted to emerge victorious yet neither side was willing to plunge the world into a nuclear holocaust in order to do so.

Paradoxically, the threat of nuclear annihilation stopped the Cold War from getting hot (U.S. State Dept., 2013).

After Kennedy was assassinated in 1963, Lyndon Johnson became President. Johnson's reelection in 1964 marked a shift in U.S. policy as American troops were sent in growing numbers to prop up the inept regimes of South Vietnam. By 1969, over 543,000 U.S. troops were fighting in Vietnam but with no defined objective beyond propping up the latest government in Saigon against attacks by the North Vietnamese army and Viet Cong guerillas (Thomas, 2014). Although Johnson ordered air campaigns against North Vietnam, he never ordered an all out ground invasion, perhaps out of fear that Russia, China or both would intervene on the side of North Vietnam.

In 1968, Johnson did not seek reelection. Richard Nixon won the Presidency promising to end the Vietnam War, which he eventually did 5 years later but only after thousands more Americans lost their lives for no good reason. Unlike the end of the Korean War 20 years earlier, the end of the Vietnam War was a complete fiasco for the U.S. military. South Vietnam did not survive as an independent country but quickly succumbed to North Vietnamese aggression. Saigon was captured in April, 1975 barely 2 years after the last U.S. ground troops withdrew from South Vietnam.

By the early 1970's, Nixon and Secretary of State Kissinger adopted a foreign policy of détente (relaxing of tensions); this was a return to the policy of the Eisenhower years but with a twist. In 1972, Nixon visited Communist China and reestablished relations (which had been severed after the Communist takeover in 1949). This gesture was largely symbolic but also designed to catch the USSR off guard. The U.S. was doing whatever it could to drive a wedge between these two powers – politically, economically, and militarily.

Afghanistan and Nicaragua – the final battlegrounds of the Cold War.

During his first term as President, Ronald Reagan used a combination of rhetoric, diplomacy, and behind the scenes deals to undermine the global influence of the Soviet Union. In one famous speech, Reagan referred to the Soviet Union as "an evil empire" and to communism as "the focus of evil in the modern world" (Kengor, 2013). Against the advice of critics and fiscal conservatives, Reagan massively increased government spending, especially in the realm of military technology. He also found ways to funnel weapons to any resistance movement dedicated to fighting communist regimes in general and the Soviets in particular.

In Afghanistan, the U.S. supplied weapons to resistance fighters called the mujahedeen via Pakistani, Saudi, and Israeli intermediaries.

As the Soviet occupation dragged on, opposition grew within the ranks of the Politburo and Soviet military, not to mention among the Russian populace. Reagan's strategy proved successful; Afghanistan became the Soviet Union's Vietnam. By 1989, the Soviets lost an estimated 13,000 troops occupying Afghanistan and withdrew their forces that same year

(Taubman, 1988). Even as Gorbachev ended Soviet support for the Najibulah regime in Kabul, he faced economic turmoil at home. In mid-1991, Soviet officers launched a coup against Gorbachev. Although the coup failed, Gorbachev resigned from office a few months later. His successor Boris Yeltsin presided over the collapse of the Soviet Union and the chaotic years that followed (Lunderstad, 2000).

Scholars continue to debate the reasons behind the Soviet invasion of Afghanistan. Some offer the traditional response that the Soviets wanted to avoid looking weak by allowing a rag tag army of guerilla fighters to overthrow their puppet government in Kabul. Others propose a more sinister motive: Brezhnev ordered the invasion of Afghanistan as a dress rehearsal for a future Soviet invasion of the entire Middle East. This scenario may seem farfetched until one recalls that the U.S. allegedly formulated a similar plan under Richard Nixon – to seize oil fields in Saudi Arabia, Kuwait, and the UAE in the event of a prolonged OPEC embargo (Alvarez, 2004).

In Nicaragua, Reagan sought to arm anti-Communist guerillas called the Contras even though Congress had passed a law limiting assistance to Nicaraguan opposition groups to humanitarian aid. Reagan's second term was overshadowed by the Iran-Contra affair, which the media first reported in 1986. The Iran-Contra Affair is often described superficially as trading arms for hostages. The Iranians released American and British hostages in exchange for weapons in their seemingly endless war against Iraq. The twist was that Colonel Oliver North used profits from these secret weapons sales to buy additional weapons, which the Central Intelligence Agency (CIA) and various mercenaries delivered to the Contras (Hershberg, 2003). In order to escape detection by the Congressional Armed Services Committee, all funds used to arm the Contras were hidden in secret bank accounts in Switzerland and Brunei (American Experience, 2013).

Reagan may or may not have known about the Iran-Contra affair, but VP Bush, as a former CIA director, was almost certainly in the loop and may even have helped orchestrate the affair. Most analysts think that Col. North and his boss Admiral Poindexter could not have set up an international network of arms dealers and secret bank accounts without CIA assistance. Although President Reagan was never directly implicated in the Iran-Contra Affair, it paralyzed his last two years in office. The outcome of the Iran-Contra Affair was anticlimactic to say the least. Admiral Poindexter resigned from the Navy, and Col. North was fined $150,000 and sentenced to perform 1,200 hours of community service (Walsh, 1993).

After the Soviet Union's official dissolution at the end of 1991, the next decade was relatively tranquil compared to the Cold War era or the years following the September 11th attacks. The U.S. acted as the dominant global superpower, intervening militarily in a few regional conflicts – Somalia in 1993; the Balkans in 1995 and Kosovo in 1999. Neither Russia nor China interfered in any of these hot spots beyond some predictable verbal criticism.

In the post-9/11 world, the most obvious indicator of a reversion to a Cold War power balance is the Russian resurgence under Vladimir Putin.

A brief military incursion into Georgia in 2008 and the annexation of Ukraine's Crimean peninsula in 2014 leave little doubt that Putin harbors an expansionist agenda. Scholars of Russian history are quick to quote Carter era national security adviser Zbigniew Brzezinski, "Without Ukraine, Russia ceases to be an empire; with Ukraine, it automatically turns into an empire." (Sengupta, 2014)

Economically, China seems like an obvious winner given its tremendous economic expansion from 1989 to present. China leads the world in rice farming, coal extraction, hydroelectric dams and aquaculture.

The downside is that China is also a world leader in massive environmental damage ranging from air pollution to soil erosion to deforestation (Diamond, 2006). More recently, India has emerged as a rising economic and military power with nuclear capabilities dating to 1972. Although traditionally a non-aligned power, India has established closer ties to the U.S. and Israel in recent years.

Nonetheless, the fundamental balance of global power has not shifted dramatically since the 1960's when China joined the U.S. and Russia as a nuclear power. No one country can hope to dominate the other two superpowers (let alone the world) in a military or even an economic sense. The U.S., Russia, and increasingly China are food self sufficient and to a large extent energy independent. Barring a catastrophe like a nuclear war, these 3 countries have vast enough populations and resource bases to dominate their respective spheres of influence for the foreseeable future. In this context, each superpower emerged victorious relative to the rest of the world.

On a deeper level, global security has declined dramatically in the post-Cold War era. Without the American and Soviet navies to patrol the world's oceans, piracy has returned with a vengeance in chaotic regions like Somalia and the Indonesian archipelago. Pirates are a minor annoyance, however, compared to the dangers of nuclear proliferation. All sides have failed to contain the spread of nuclear weapons in the post-CW era to pariah or rogue states like Iran and North Korea; on this count no country can claim victory. Although China has virtual control of North Korea's government, Beijing cannot predict, let alone control, Kim Jong Un's agenda. Nuclear weapons in the hands of a deranged dictator are clearly a recipe for global disaster.

Iran represents the opposite side of that coin. Since the death of the Ayatollah Khomeini in 1989, the Islamic Republic has been controlled by a shadowy leader known as Ayatollah Ali Khameini, who approves the appointment of a President during each election cycle (PBS, 2001). Outside observers disagree as to how much real power the Iranian President wields compared to the Supreme Ayatollah. Most intelligence agencies outside Iran are convinced, however, that the Iranian regime is attempting to build nuclear warheads under the guise of a civilian nuclear power program. The most obvious target of an Iranian nuclear attack would be Israel, but with ever longer ranges on its rockets (as well as terrorist proxies willing to die in suicide attacks), Iran could conceivably target any country in the world with a "dirty" bomb. Whether Israel launches a unilateral attack against Iran's nuclear facilities (as it did against Iraq's reactor at Osirak in 1981) is anyone's guess.

In conclusion, the fall of the Berlin Wall and collapse of the Soviet Union marked the end of the Cold War. These events were clear victories for the U.S., at least in the short term. The so called peace dividend with the U.S. enjoying sole superpower status lasted about a decade from 1992 until the 9/11 attacks of 2001.

Today Vladimir Putin is the de facto czar of Russia. By methodically reasserting Russian control over the former soviet republics, he is slowly but inexorably rebuilding the Soviet Union. Neither the U.S. nor the NATO alliance appears willing or able to challenge Russian resurgence as long as Putin remains in power. Meanwhile, China is on a trajectory to become the dominant global economic superpower by the middle of the 21st century. Neither the U.S. nor Russia won this aspect of the Cold War.

It is unclear whether any nation is powerful enough to manage global security unilaterally, at least in the realm of nuclear proliferation. During the Cold War era, fewer than 10 nations were known to possess nuclear weapons (CNN, 2013). Today, North Korea has joined the ranks of nuclear powers, and Iran is intent on doing so. These countries along with Pakistan are the most likely states to implode into chaos; without concerted action on the part of a global coalition, the nightmare scenario of nuclear weapons falling into the hands of terrorists could become a reality. And in that case, there will be no winners.

Essay 122: Was the League of Nations a Failure?

The short answer is yes, but this raises an obvious question: Why did the League of Nations fail? To answer this question adequately, it is first necessary to understand the historical context in which the League of Nations was founded. Next it is important to analyze examples of the League's role in conflict mediation in the two decades between World War I and World War II. Finally, given that the League of Nations did not prevent the outbreak of a second world war, it is critical to distinguish the proximate and ultimate causes of the League's failure to accomplish its intended mission.

The League of Nations was established in the aftermath of World War I, or the Great War as contemporary historians called it. Four years of stalemate and carnage left an estimated 20 million soldiers and civilians dead and devastated much of central and Eastern Europe. At the Paris Peace Conference in 1918, U.S. President Woodrow Wilson proposed the creation of an international body to promote world peace, combat imperialist aggression and resolve future conflicts diplomatically whenever possible. The League of Nations represented the fourteenth and final point of Wilson's framework for peace in post-war Europe. In Wilson's words:

XIV. A general association of nations must be formed under specific covenants for the purpose of affording mutual guarantees of political independence and territorial integrity to great and small states alike.

Wilson returned to the U.S. in 1919 resolved to convince Congress and the American people of the importance of supporting the League of Nations in its quest to preserve peace in Europe and beyond. When isolationist elements in Congress blocked ratification of the Treaty of Versailles, Wilson engaged in a nationwide speaking tour, traveling throughout the country in a relentless effort to muster popular support for U.S. membership in the League.

Despite Wilson's monumental efforts, Congress and the American public remained unconvinced that U.S. involvement in European affairs or membership in the League of Nations was important or even desirable. In the end, the stress of non-stop travel and speaking engagements wrecked Wilson's health. In September 1919 Wilson suffered a debilitating stroke, effectively ending his presidency and shattering any chance of the U.S. joining the League of Nations.

Just over a year later, Warren Harding won the 1920 presidential election. Campaigning on a platform of a "return to normalcy" Harding's presidency marked the start of a two decades long U.S. retreat into isolationism.

Postwar Europe: a peace built on quicksand

Wilson presented his 14 points at the Paris Peace Conference in the hopes that European leaders would cooperate in nurturing harmony (or at least coexistence) as Europe recovered from the horrific war. Much like Wilson's dream of the U.S. playing a key role in the League of Nations, however, this vision of European cooperation was dead on arrival.

When asked why the U.S. entered WWI on the allied side (instead of remaining neutral) Wilson responded that the goal of the U.S. and its allies was to "make the world safe for democracy."(65th Congress) Britain, France and several countries that had fought Germany for 4 years had other ideas. They were more interested in punishing Germany economically than they were in supporting the fledgling Weimar Republic. Considering the half hearted support of the Allies, not to mention the apathy of large segments of German society, it is little wonder democracy never took root in postwar Germany.

The prospects for democracy in Russia and Eastern Europe after WWI were even bleaker than in Germany. In 1917, Russia erupted into revolution. Czar Nicholas II was deposed and exiled to Siberia where he and the royal family were eventually killed. A provisional government led by Alexander Kerensky held power in Russia for less than a year before it was overthrown by Lenin and the Bolsheviks. The Bolshevik takeover of 1917 marked the start of a protracted civil war in Russia. Although the Allies and Japan supported anti-Communist militants (the 'Whites'), Communist 'Red' forces prevailed and established the Soviet Union in 1922.

While the Soviet Union was mainly interested in quashing internal dissent during its first decade of existence, Josef Stalin (who succeeded Lenin in 1924) expanded the reach of Soviet power in the 1930's. The Soviets eventually signed a non-aggression pact with Nazi Germany in 1939 just prior to the German invasion of Poland. As WWII erupted, Soviet forces captured the Baltic republics of Estonia, Latvia and Lithuania in June 1940.

Fascism appeared on the international scene with Mussolini's rise to power in Italy in 1922 but gained momentum only after Hitler and the Nazis consolidated power in Germany in 1933. While historians still debate the root causes of the rise of the Third Reich, it is no coincidence that the Nazis took power only after the stock market crash of 1929 plunged the U.S. and much of the world into the Great Depression. Any enthusiasm that Germans felt for democracy in the 1920's abruptly ended as the Weimar Republic unraveled into bankruptcy and impotence in 1930.

In 1936, Spain (already beleaguered politically and economically) erupted into civil war. Although the Soviet Union as well as volunteer forces from League of Nations member states fought to preserve the Loyalist government, fascist forces, commanded by Francisco Franco and armed by Nazi Germany, ultimately prevailed. Most historians agree that the Spanish Civil War served as a test run for the Nazi invasion of Europe in WWII.

The Rise of Imperial Japan

If the League of Nations was faltering in Europe, its weakness as an organization was blatantly obvious in East Asia. In Japan, nominally a member of the WWI alliance, a militarist faction led by Hideki Tojo gained the upper hand in the Diet (Japanese Parliament) by 1926. In 1930, Japan's civilian government collapsed following the assassination of Prime Minister Osachi at the hands of an ultra-nationalist.

In 1931, Japanese forces (which had occupied Korea since 1910) used the peninsula to launch an invasion of Manchuria. China's nationalist government, already plagued by internal dissent and worn down by fighting local warlords in the 1920's, could not withstand the Japanese onslaught. The League of Nations responded predictably by condemning Japanese aggression; however, the League lacked any military power to force Japan to withdraw from China. Japan's response to the League's criticism was also predictable. The militarist government withdrew Japan from the League of Nations and went on to become an ally of Fascist Italy and Nazi Germany.

Success as a Humanitarian Organization

The League of Nations could point to a few successes on the diplomatic and humanitarian fronts in the 1920's. These included mediating conflicts in the Balkans (Yugoslavia and Albania) as well as between Greece and Bulgaria. The League also diffused tension between Turkey and Iraq as the two countries contended over the city of Mosul. The League's greatest humanitarian success was repatriating over 400,000 prisoners of war and other refugees in the aftermath of WWI. While these successes were commendable, they invariably involved countries that lacked the ability to project military power.

This became a recurrent theme in the League's history: the organization could win small battles but had no power to stop the global march to war.

Proximate reasons for the League's failure:

The U.S. retreated into isolationism until the Japanese attack on Pearl Harbor in 1941.

Britain and France were more concerned with running colonial empires than with combating the rise of fascism in Europe.

Japan's invasion of Manchuria in 1931 exposed the League of Nations as an organization strong on talk but weak on action.

Italy's invasion of Ethiopia in 1936 and the outbreak of the Spanish Civil War the same year revealed the League of Nations to be a powerless force in opposing fascist aggression.

The Munich conference in 1938 signaled Europe's headlong slide into a second world war. With Germany's annexation of Czechoslovakia in March 1939, the League became irrelevant.

Ultimate Reasons the League of Nations failed

The conflicts leading to WWI in Europe remained unresolved after 1918. If WWI was the war to end all wars, the Treaty of Versailles was the peace to end all peace.

With or without the League of Nations, postwar Europe lacked the collective will to nurture democracy in the former empires of Germany and Austria-Hungary or challenge the fascist threat posed by Italy and later by Nazi Germany.

Post-WWI Germany rejected democracy as a political system and instead embraced totalitarianism in the form of a Nazi government bent on world domination.

No country in the WWI alliance was prepared to assume a leadership position in the League of Nations nor were their leaders willing to support the League's peacekeeping efforts militarily.

An international peacekeeping organization without the means to combat aggression militarily amounts to little more than a university debate club. Not surprisingly, this fundamental weakness plagues the League of Nations' modern day counterpart the United Nations.

Essay 123: The Yuan Dynasty and Medieval China

The Yuan Dynasty was the title that later Chinese historians applied to the 150-year era of Mongol rule over China. This period began in the year 1215, with the surrender of Peking to the army of Genghis Khan, and lasted until 1368, when widespread revolts brought the Ming dynasty to power. During the one-and-a-half centuries of the Yuan dynasty, the Mongol rulers of China gradually adopted Chinese customs and culture, especially in the realm of court life. In governing China, they kept the civil service and local bureaucracies intact. In foreign policy, the Mongol leadership continued to expand the borders of its vast empire by conquest and diplomacy.

The peak of the Yuan dynasty came under Kublai Khan, grandson of Genghis Khan, who reigned from 1260 to 1294. A great deal of our knowledge of China under Kublai Khan comes from the book The Travels of Marco Polo, dictated by Polo several years after his return from China around 1295. Aside from Marco Polo's father and uncle, few other medieval Europeans traveled the entire length of the Silk Road at the height of the Mongol empire.

Marco Polo lived in China for 20 years, befriended Kublai Khan, and was eventually appointed governor of the city of Canton (then called Guangzhou). According to Polo's account, Kublai Khan spent some of each year traveling throughout the Chinese region of his empire but spent most summers at his palace at Xanadu in Mongolia. Throughout China, travel was completely safe; most of the Mongol empire had a virtual absence of banditry. The Khan also supported the arts theater and the maintenance of churches, mosques, synagogues, and temples throughout Asia.

During his long reign, Kublai Khan also continued the Mongol policy of bloodless conquest in return for tax or tribute. Southern China, still under the rule of the Sung dynasty, surrendered without a fight in 1260. Some years later, the kingdom of Korea was peacefully annexed by the Mongols. A few coastal areas in India also became tributary states, presumably without a fight. Conversely, any state that refused this offer could expect no mercy from the Mongol army.

Two years prior to Kublai's reign, Mongol forces destroyed Baghdad in 1258 when its caliph refused to become a vassal. In 1283, the Burmese state of Pagan fought the Mongols to a draw but later paid a large tribute to retain its independence.

During the entire 13th century, there were only a few instances of successful military resistance to Mongol rule. The first came in 1260 when the Egyptians hired enough Turkish and Mongol mercenaries (collectively called Mameluks) to halt the Mongol advance at Lake Tiberias in modern-day Israel. Closer to home, Japan warded off two Mongol invasion forces in 1274 and again in 1281. According to Marco Polo, Kublai Khan drew up plans for a third invasion fleet, but a total lack of enthusiasm effectively ended the project before it began.

From the perspective of Western historians, the Yuan dynasty was the one time in history when China completely lost its political independence to foreign conquerors. Nonetheless, the basic character of life stayed the same for most Chinese, who remained peasant farmers or local merchants. As memories of the Mongol conquest faded, most no doubt regarded the Yuan era as one dynasty overthrowing its predecessor and, in time, being overthrown itself. At some point in history, the Chinese adopted a now famous expression, "China always conquers its conquerors."

Epilogue

It's difficult to wrap up a book that was 30 years in the making. Each of these essays could be expanded into books of their own. Instead of writing something forgettable, I will use this last page as the segue to my next book entitled "I'll Be Better". As with any good autobiography, it has humor, tragedy, philosophy and all that jazz.

Dear reader, you can probably relate to this - a chapter of life has ended; you find yourself at a crossroads; events seem overwhelming, making decisions is difficult, and some encouraging words would be nice.

So let me paraphrase a lesson from a Chabad calendar I read 20 years ago. It's really beautiful. And I will for sure include it in I'll Be Better.

"If you did not need this world, and this world did not need you, G-d would never have brought you here in the first place. You say you cannot see a plan. Should it surprise you that a finite being cannot fathom the wisdom of an Infinite Creator? You say you feel stuck, like you are spinning your wheels but going nowhere. Some trees bear fruit after a few years. Others like carob trees can take 70 years or more to bear fruit. But carobs bear fruit for centuries. As long as your dreams are good, never give up on them. Plant and sow, the fruits will come."

Now look in the mirror, and no matter how you feel, keep in mind that you need this world, and this world needs you. Now say it out loud, "I will never be perfect. I'll be better!"

And with that...farewell until next time.

References

American Entry into WWI https://history.state.gov/milestones/1914-1920/wwi

Constitution Center, the 22nd Amendment http://constitutioncenter.org/constitution/the-amendments/amendment-22-presidential-term-limits

Cruise of the Great White Fleet

http://www.history.navy.mil/library/online/gwf_cruise.htm

Military History, 2014 http://militaryhistory.about.com/od/battleswars1900s/p/greatwhitefleet.htm

Election of 1876

http://www.fofweb.com/History/MainPrintPage.asp?iPin=DACH0075&DataType=AmericanHistory&WinType=Free

http://presidentialcampaignselectionsreference.wordpress.com/overviews/19th-century/1876-overview/

Facebook.com, Barack Obama, 2014 https://www.facebook.com/barackobama

FDR: The New Deal

http://rooseveltinstitute.org/policy-and-ideasroosevelt-historyfdr/new-deal

The Impeachment Trial of President William Clinton

http://law2.umkc.edu/faculty/projects/ftrials/clinton/clintontrialaccount.html

The Iran-Contra Affair

http://www.pbs.org/wgbh/americanexperience/features/general-article/reagan-iran/

New York Times Archives, Taubman, P. 'Soviet Lists Afghan War Toll' 1988

http://www.nytimes.com/1988/05/26/world/soviet-lists-afghan-war-toll-13310-dead-35478-wounded.html

Panamanian Independence

http://www.history.com/this-day-in-history/panama-declares-independence

Sinking of the Lusitania   http://history1900s.about.com/cs/worldwari/p/lusitania.htm

Theodore Roosevelt, Key Events http://millercenter.org/president/roosevelt/key-events

U.S. Census Bureau, 2000

Census of 1860 https://www.census.gov/history/www/through_the_decades/overview/1860.html

Census of 1900 http://www.census.gov/population/estimates/nation/popclockest.txt

U.S. National Archives, The Zimmerman Telegram
http://www.archives.gov/education/lessons/zimmermann/

Walsh, Iran / Contra Report, U.S. v. Oliver L. North http://fas.org/irp/offdocs/walsh/chap_02.htm

Watergate Scandal, 2014

http://watergate.info/

http://www.u-s-history.com/pages/h1791.html

http://www.pbs.org/johngardner/chapters/6c.html

References and Further Reading

Essay	Title	Reference
1	Alpha-1 Antitrypsin Deficiency	https://www.sciencedirect.com/science/article/pii/S0954611111000515
2	Alzheimer's Disease	https://www.cdc.gov/aging/aginginfo/alzheimers.htm
3	Alcoholism	1. Alcoholism http://www.nlm.nih.gov/medlineplus/alcoholism.html 2. Liver Cancer / HCC http://www.ncbi.nlm.nih.gov/pubmedhealth/PMH0001325/ 3. Endocarditis http://www.ncbi.nlm.nih.gov/pubmedhealth/PMH0002088/ 4. Wernicke-Korsakoff Syndrome http://www.ncbi.nlm.nih.gov/pubmedhealth/PMH0001776/ 5. Alcoholic Cardiomyopathy http://emedicine.medscape.com/article/152379-overview 6. Delirium Tremens http://www.ncbi.nlm.nih.gov/pubmedhealth/PMH0001771/
4	Amino Acid Metabolism	L. Stryer's Biochemistry 1988: Protein Structure and Function; Amino Acid Degradation and the Urea Cycle; Biosynthesis of Amino Acids and Heme
5	Amygdala	1. Amygdala http://biology.about.com/od/anatomy/p/Amygdala.htm 2. Kluver-Bucy Syndrome http://www.dana.org/Cerebrum/Default.aspx?id=39212 3. Papez Circuit http://www.encyclo.co.uk/2013/define/Papez%20circuit
6	Anti-arrhythmics	1. Amiodarone http://www.nlm.nih.gov/medlineplus/druginfo/meds/a687009.html 2. CAST https://www.ahajournals.org/doi/full/10.1161/01.cir.91.1.245

		3. Procainamide http://reference.medscape.com/drug/procanbid-pronestyl-procainamide-342306#4
7	Antibiotics	https://www.orthobullets.com/basic-science/9059/antibiotic-classification-and-mechanism
8	Arthritis	Arthritis Overview https://www.ncbi.nlm.nih.gov/pmc/articles/PMC8572231/
9	Autopsy	https://oregon.public.law/statutes/ors_97.170
10	Biological Basis of Memory	1. Donald Hebb and Synaptic Plasticity http://hebb.mit.edu/courses/8.515/lecture2/sld014.htm 2. The Case of H.M. http://www.nytimes.com/2008/12/05/us/05hm.html?_r=1& 3. Morris Water Maze http://www.watermaze.org/
11	Bioluminescent Fungi	http://warnell.forestry.uga.edu/service/library/index.php3?docID=173
12	Human Bones Aging	https://www.ncbi.nlm.nih.gov/pmc/articles/PMC2991386/
13	Brain Osmolality and Water Balance	1. Aquaporins http://www.ionchannels.org/showabstract.php?pmid=7859914 2. Osmolality Blood Test http://www.nlm.nih.gov/medlineplus/ency/article/003463.htm 3. SIADH http://chp.staywellsolutionsonline.com/Library/Encyclopedia/90,P01974 4. Supraoptic Nucleus and Vasopressin http://press.endocrine.org/doi/full/10.1210/jcem.84.12.6187
14	Caffeine	https://medlineplus.gov/caffeine.html
15	Carbon-14 Dating	1. Isotopes – Radioactive Decay http://www.colorado.edu/physics/2000/isotopes/radioactive_decay.html 2. Uranium Radiation Properties http://www.wise-uranium.org/rup.html

16	Cardiomegaly	Tetralogy of Fallot http://www.ncbi.nlm.nih.gov/pubmedhealth/PMH0002534/
17	Cell Junctions	http://users.rcn.com/jkimball.ma.ultranet/BiologyPages/J/Junctions.html
18	Cholesterol	1. ExRx.net Testing LDL and HDL http://www.exrx.net/Testing/LDL&HDL.html 2. L. Stryer, *Biochemistry 3rd edition* 1988, Chapters 12 and 23, Introduction to Biological Membranes and Biosynthesis of Membrane Lipids and Steroid Hormones
19	Chromosomal Aneuploidy	1. Trisomy 13 Patau Syndrome http://www.ncbi.nlm.nih.gov/pubmedhealth/PMH0002625/ 2. Pallister-Killian Syndrome http://ghr.nlm.nih.gov/condition/pallister-killian-mosaic-syndrome
20	Climate Change Over the Last 1000 Years	1. Medieval Warm Period http://books.google.com/books?id=z-BWE4iCrfYC&pg=PA134#v=onepage&q&f=false 2. Little Ice Age http://www.meteo.psu.edu/holocene/public_html/shared/articles/littleiceage.pdf 3. Mt. Tambora Eruption http://www.streetdirectory.com/travel_guide/indonesia/indonesiavolcanoes/info-70-mount_tambora_eruption.php 4. Krakatoa http://history1800s.about.com/od/thegildedage/a/krakatoa.htm 5. Maunder Minimum http://science.jrank.org/pages/4184/Maunder-Minimum.html 6. Thermohaline Convection http://www.killerinourmidst.com/THC.html 7. Angkor Wat http://www.holiday-in-angkor-wat.com/angkor-wat-history.html 8. Global Temperatures http://ete.cet.edu/gcc/?/resourcecenter/viewResource/3/
21	Colon Cancer	1. Mayer RJ, Harrison's Principles of Internal Medicine 14th edition (1998) Gastrointestinal Tract Cancer, Ch. 92, 568-578. 2. http://www.mayoclinic.com/health/colon-cancer/DS00035/DSECTION=risk-factors

22	Causes of Anemia	1. https://www.mayoclinic.org/diseases-conditions/iron-deficiency-anemia/symptoms-causes/syc-20355034 2. https://www.nhs.uk/conditions/vitamin-b12-or-folate-deficiency-anaemia/
23	Common Myths of Medical School Acceptance	Real Life Experience ;-)
24	Constellations – Telling Time	http://astronomycentral.co.uk/northern-hemisphere-constellation-maps/ http://www.nightskyinfo.com/maps_images/html/sky_map_south.htm
25	Constellations: Amateur Star Gazers	http://starryskies.com/The_sky/constellations/
26	Cassiopeia the Queen	Cassiopeia, the Queen http://www.ianridpath.com/startales/cassiopeia.htm
27	Copper Toxicity in Sheep	https://www.icliniq.com/articles/healthy-living-wellness-and-prevention/copper-toxicity
28	Cosmology- a Brief History	Copernicus http://scienceworld.wolfram.com/biography/Copernicus.html Galileo http://plato.stanford.edu/entries/galileo/ Newton http://inventors.about.com/library/inventors/blnewton.htm Einstein http://www.fearofphysics.com/Relativity/relativity.html
29	Courses to take before Pharmacy School	Real Life Experience ;-)
30	Crohn's Disease / IBD	http://www.ncbi.nlm.nih.gov/pubmedhealth/PMH0001295/
31	Cygnus the Swan	http://www.constellation-guide.com/constellation-list/cygnus-constellation/

32	Destinations in Mexico and Caribbean	Family trips to everywhere except Mexico City. Back in 2002, the Mexican Riviera was an enjoyable vacation. In 2024…bring a bulletproof vest ☹ http://www.google.com/search?q=chichen+itza
33	Diabetes Mellitus	Harrison's Internal Medicine, and scrape the icing off the cupcake.
34	Diagnosis of COPD	GOLD Report 2023 https://goldcopd.org/2023-gold-report-2/
35	Diet and Metabolism	Cushing's Disease: http://www.ncbi.nlm.nih.gov/pubmedhealth/PMH0001388/ Hashimoto's Thyroiditis: http://www.mayoclinic.com/health/hashimotos-disease/DS00567
36	Different Strokes	SAH http://www.ncbi.nlm.nih.gov/pubmedhealth/PMH0001740/ ICA Thrombosis http://www.mayfieldclinic.com/PE-CarotidStenosis.htm Status Migrainosus http://www.aafp.org/afp/2000/1101/p2145.html
37	Dreyfus Affair	http://www.jewishvirtuallibrary.org/jsource/anti-semitism/Dreyfus.html http://www.jewishvirtuallibrary.org/jsource/biography/Herzl.html
38	Economics 101	Undergraduate memories
39	Edwin Hubble Ho	https://www.cfa.harvard.edu/~dfabricant/huchra/hubble/
40	Election 2016	Personal observations and analysis
41	Ergot Alkaloids	http://botit.botany.wisc.edu/toms_fungi/oct99.html

42	Eukaryotic Cells	1. http://library.thinkquest.org/C004535/eukaryotic_cells.html
		2. http://straightlab.stanford.edu/kinetochore.html
		3. Chromosomes and expression mechanisms Current Opinion in Genetics & Development Volume 15, Issue 2, April 2005, Pages 153–162
		http://www.sciencedirect.com/science/article/pii/S0959437X05000195
		4. The Endosymbiotic Hypothesis, https://endosymbiotichypothesis.wordpress.com/history-the-formation-of-the-endosymbiotic-hypothesis/
43	Evolution of the American Presidency	If you can remember the Iran Contra Affair, it was Bush Sr. all along. Otherwise, watch the History Channel, YouTube, and the outcome of the 2024 Election
44	The Human Eye	Glaucoma http://www.ncbi.nlm.nih.gov/pubmedhealth/PMH0002587/
		LHON http://www.livestrong.com/article/162285-list-of-mitochondrial-diseases/
		Retina http://www.nlm.nih.gov/medlineplus/ency/article/002291.htm
45	Fall Foliage	Road trips and shows on PBS
		Patapsco State Park http://www.dnr.state.md.us/publiclands/central/patapsco.asp
		Appalachian Trail http://www.google.com/search?q=maryland+appalachian+trail
		Shenandoah National Park http://www.nps.gov/shen/index.htm
46	Fat	L. Stryer, Biochemistry, 3rd ed., 1988, Ch.12 Introduction to Biological Membranes, pp. 283-314; Ch. 20 Fatty Acid Metabolism, pp. 469-494; Ch. 23 Biosynthesis of Membrane Lipids and Steroid Hormones pp. 547-574, and Ch. 38 Hormone Action, pp. 975-1004.
47	Fish Anatomy	L. Stryer Biochemistry, 1988 (the electric eel paragraph)
48	Five Misdiagnosed Diseases	SLE http://www.ncbi.nlm.nih.gov/pubmedhealth/PMH0001471/
		Sarcoidosis http://www.ncbi.nlm.nih.gov/pubmedhealth/PMH0001140/
		MS http://www.ncbi.nlm.nih.gov/pubmedhealth/PMH0001747/

		Pheochromocytoma http://www.ncbi.nlm.nih.gov/pubmedhealth/PMH0001380/ Atypical MI http://www.ncbi.nlm.nih.gov/pubmedhealth/PMH0001246/
49	Fluorescent Proteins	https://www.microscopyu.com/techniques/fluorescence/introduction-to-fluorescent-proteins
50	Four Classes of Biomolecules	L. Stryer but definitely check out Voet & Voet Biochemistry
51	Geography of the Atlantic Ocean	Atlantic Ocean http://www.eoearth.org/article/Atlantic_Ocean Mid-Atlantic Ridge http://www.britannica.com/EBchecked/topic/380800/Mid-Atlantic-Ridge Sargasso Sea http://www.gobi.org/Our Work/rare-2
52	Geography and I.R.	Guns, Germs and Steel / Logical Analysis
53	Geological Processes in the Earth's Interior	Earth's Magnetic Field http://science.nasa.gov/science-news/science-at-nasa/2003/29dec_magneticfield/eld Ring of Fire http://www.enchantedlearning.com/subjects/volcano/ringoffire/
54	German Military Blunders WWI	1. PBS: The Great War http://www.pbs.org/greatwar/ 2. Schlieffen Plan http://www.historylearningsite.co.uk/schlieffen_plan.htm 3. Sinking of the Lusitania http://history1900s.about.com/cs/worldwari/p/lusitania.htm 4. Zimmermann Telegram http://www.archives.gov/education/lessons/zimmermann/
55	Glucose and Sodium Homeostasis	Images courtesy of: http://www.lifeextension.com/magazine/2004/6/report_diabetes/page-02?p=1 https://www.britannica.com/science/renin-angiotensin-system http://ajprenal.physiology.org/content/277/3/F319 http://www.nbs.csudh.edu/chemistry/faculty/nsturm/CHE452/22_Ren AngioAldoANP18.htm

		Katzung B, Trevors J and Masters SB (2009) Basic and Clinical Pharmacology 11th ed. McGraw Hill. MacWilliam L (2004) Diabetes, *Life Extension Magazine* http://www.lifeextension.com/magazine/2004/6/report_diabetes/page-02?p=1 Sturm N. (2017) Renin / Angiotensin / Aldosterone / ANP http://www.nbs.csudh.edu/chemistry/faculty/nsturm/CHE452/22_Ren AngioAldoANP18.htm Verrey F (1999) American Journal of Physiology - Renal Physiology **Vol.** 277 **no.** 3, F319-F327
56	Hashimoto's Thyroiditis	https://www.thyroid.org/hashimotos-thyroiditis/
57	Heart Healthy Foods	Fish and Fruits good; Refined Sugar and Red Meat bad. When in doubt, order a large bowl of steam.
58	Heat Stroke	http://www.nlm.nih.gov/medlineplus/ency/article/000356.htm
59	Honeybees	Colony Collapse Disorder, 2013 http://www.ars.usda.gov/News/docs.htm?docid=15572
60	How Do Neurotoxins Work?	Some fast, some slow ☺ Black widows http://www.emedicinehealth.com/black_widow_spider_bite/article_em.htm Cobras http://cobras.org/cob_3.htm Puffer fish http://www.sma.org.sg/smj/4502/4502a2.pdf
61	How Do Steroid Hormones Work?	By turning various genes on and off. Glucocorticoids http://www.nlm.nih.gov/medlineplus/steroids.html Mineralocorticoids http://www.vivo.colostate.edu/hbooks/pathphys/endocrine/adrenal/mineralo.html Addisonian Crisis http://www.ncbi.nlm.nih.gov/pubmedhealth/PMH0001397/
62	How Space and Time	1. http://library.thinkquest.org/06aug/02088/newton.htm

	Became Spacetime	2. http://www.astro.cornell.edu/academics/courses/astro201/merc_adv.htm

3. http://www.goodreads.com/book/show/1405966.A_Journey_Into_Gravity_And_Spacetime

4. http://rsta.royalsocietypublishing.org/content/220/571-581/291.full.pdf

5. http://news.discovery.com/human/psychology/time-dilation-einstein-relativity.htm

6. http://scienceblogs.com/principles/2013/04/29/what-is-time/ |
| 63 | Human Circulatory System | http://www.brianmac.co.uk/physiolc.htm

http://www.cvphysiology.com/Blood Pressure/BP015.htm |
| 64 | Human Immune System | Janeway Immunology 1995

Abbas Immunology Review 1998

Analogies to the U.S. military and ninjas are all me |
| 65 | From Hunter Gatherers to Farmers | Jared Diamond, *Guns, Germs and Steel*, 1997

Available at Barnes and Noble, Amazon, and the public library ☺

http://www.pbs.org/gunsgermssteel/ |
| 66 | Huntington's Chorea | http://www.ncbi.nlm.nih.gov/pubmedhealth/PMH0001775/ |
| 67 | Primary/Essential Hypertension | http://www.uptodate.com/contents/lacunar-infarcts |
| 68 | Secondary HTN | Conn Tumor https://www.adrenal.com/conn-syndrome/primary-hyperaldosteronism

Cushing's Syndrome https://www.niddk.nih.gov/health-information/endocrine-diseases/cushings-syndrome |
| 69 | HTN – Non-Pharmacolo | Don't smoke / Don't drink / Exercise / Vegetables are our friends |

	gical Treatments	
70	Incredible Insulin	https://www.uptodate.com/contents/insulin-action
71	International Security and Global Balance of Power	National Geographic http://news.nationalgeographic.com/news/energy/2010/11/101109-peak-oil-iea-world-energy-outlook/
72	Judaism – Apocryphal Literature	Jewish Virtual Library https://www.jewishvirtuallibrary.org/the-apocrypha-and-pseudepigrapha
73	Judaism – The Patriarchs	Book of Genesis, preferably with Rashi commentary
74	Judaism – The Star of David	Chaim Potok, *Wanderings – A History of the Jews*, 1978
75	Leo the Lion	https://www.britannica.com/place/Leo-constellation http://www.astro.wisc.edu/~dolan/constellations/constellations/Leo.html
76	Leprosy	http://www.nlm.nih.gov/medlineplus/ency/article/001347.htm Clofazamine HSRA research http://boards.medscape.com/forums/
77	Leukemia	Imatinib (Gleevec) http://www.gleevec.com/index
78	Long Term Potentiation	**Retrograde messenger** http://www.pa.ibf.cnr.it/personale/migliore/PDF/biosys40.pdf
79	Lung Cancer	Harrison's Textbook of Internal Medicine https://www.cancer.gov/types/lung
80	Macronutrients	Lubert Stryer's *Biochemistry* 3rd ed. (1988) Ch. 26, Integration of Metabolism, pp. 639-644 http://www.hsph.harvard.edu/nutritionsource/what-should-you-eat/carbohydrates/

81	Melatonin	Melatonin, University of Maryland http://www.umm.edu/altmed/articles/melatonin-000315.htm
		WebMD http://www.webmd.com/drugs/drug-87797
		Melatonex Melatonex+Oral.aspx?drugid=87797&drugname=Melatonex+Oral
82	Messier Marathon	https://freestarcharts.com/messier
83	Mitochondria	http://www.cellsalive.com/cells/mitochon.htm
		http://www.sciencedirect.com/science/article/pii/0960076092903053
84	Neuroblastoma	http://www.ncbi.nlm.nih.gov/pubmedhealth/PMH0002381/
85	Neurofibromatosis	http://ghr.nlm.nih.gov/condition/neurofibromatosis-type-1
86	Neuroscience Predictions	Personal ideas
87	Neurotransmitters	http://www.brainexplorer.org/neurological_control/Neurological_Neurotransmitters.shtml
88	Oceanic Heat Content	Milankovitch cycles http://www.sciencecourseware.org/eec/GlobalWarming/Tutorials/Milankovitch/
89	Orion the Hunter	http://scienceblogs.com/startswithabang/2009/12/27/weekend-diversion-orion/
90	Parkinson's Disease	http://www.ncbi.nlm.nih.gov/pubmedhealth/PMH0001762/
91	Patau Syndrome	http://www.ncbi.nlm.nih.gov/pubmedhealth/PMH0002625/
92	Philadelphia Chromosome	t9:22 Translocation https://www.ncbi.nlm.nih.gov/pmc/articles/PMC4896164/
93	Pituitary Functions	http://www.nlm.nih.gov/medlineplus/ency/anatomyvideos/000099.htm
		Aquaporins http://aquaporins.org/peter.htm

#	Topic	Links
94	Pituitary Tumors	Prolactinomas http://www.ncbi.nlm.nih.gov/pubmedhealth/PMH0001377/ http://www.skullbaseinstitute.com/nonfunctioning_pituitary.htm
95	Platelets	http://vannellehekeleh.blogspot.com/2011/01/bila-la-nk-jadi-rajin-ni.html#!/2011/01/bila-la-nk-jadi-rajin-ni.html
96	Pneumonia	CAP http://www.ncbi.nlm.nih.gov/pubmedhealth/PMH0001200/ Nosocomial PNA http://www.ncbi.nlm.nih.gov/pubmedhealth/PMH0001201/
97	Portal Hypertension	http://my.clevelandclinic.org/disorders/portal_hypertension/hic_portal_hypertension.aspx Visceral Leishmaniasis – Keep the pentavalent antimony handy ;-) http://www.ncbi.nlm.nih.gov/pubmedhealth/PMH0002362/
98	Proteases	Voet & Voet Serine Proteases http://biochem.wustl.edu/~protease/ser_pro_overview.html
99	Quetiapine	http://www.ncbi.nlm.nih.gov/pubmedhealth/PMH0001030/
100	Rainforests	Emergent Layer http://mset.rst2.edu/portfolios/d/dispenza_l/web_design/rainforests_final/emergent_layer/emergent_layer.htm Flooded Forest http://www.nationalgeographic.com/wildworld/profiles/g200/g147.html
101	RLQ Pain	Gallstone Ileus http://www.unboundedmedicine.com/2006/08/03/gallstone-ileus/ http://www.ncbi.nlm.nih.gov/pubmedhealth/PMH0001956/ Intestinal Adhesions http://digestive.niddk.nih.gov/ddiseases/pubs/intestinaladhesions/ http://digestive.niddk.nih.gov/ddiseases/pubs/anatomiccolon/ Intussusception http://kidshealth.org/parent/system/surgical/intussusception.html Small Bowel Tumor http://www.ddc.musc.edu/public/problems/diseases/smallBowel/tumors.cfm http://www.ncbi.nlm.nih.gov/pubmedhealth/PMH0002366/

		IBD http://www.medicinenet.com/inflammatory_bowel_disease_intestinal_problems/page4.htm
102	Risk Factors CVA	http://www.medicinenet.com/stroke/article.htm
103	Rock Formation in Geology	Igneous Rocks http://www.fi.edu/fellows/fellow1/oct98/create/igneous.htm Sedimentary http://geology.com/rocks/sedimentary-rocks.shtml Metamorphic http://geology.about.com/cs/basics_roxmin/a/aa011804c.htm
104	Seven Wonders of Antiquity	http://www.unmuseum.org/wonders.htm
105	Sickle Cell Anemia	Sickle Cell Anemia http://www.nhlbi.nih.gov/health/health-topics/topics/sca/treatment.html
106	Soil Quality	The wisdom of Cousin Alan (RIP Joey McFuzz)
107	Sports Injuries	1. TV commercials / NFL games 2. Personal experience
108	Systemic Lupus Erythematosus	https://emedicine.medscape.com/article/332244-overview
109	Stages of CKD	http://www.ncbi.nlm.nih.gov/pubmedhealth/PMH0001503/
110	Staphylococcus: Rise of Superbugs	http://www.textbookofbacteriology.net/staph.html
111	Sun Moon and Tides	Sun http://nineplanets.org/sol.html Moon http://nssdc.gsfc.nasa.gov/planetary/factsheet/moonfact.html NOAA https://tidesandcurrents.noaa.gov/
112	Temperature Control in the Human Body	http://onlinelibrary.wiley.com/doi/10.1038/sj.bjp.0705655/pdf

113	Legacy of the 30 Years War	Professor Thomas Munck, can you assign this essay a B+ or higher in exchange for a free copy of this book? Please? Ottoman Siege of Vienna www.wien-**vienna**.com/**vienna**1529.php What Happened in the 30 Years War? *www.economist.com/blogs/.../01/economist-explains-5*
114	Tumors of the CNS	https://www.cancer.gov/types/brain/hp
115	Ursa Major the Great Bear	http://starryskies.com/The_sky/constellations/ursa_major.html M97, the Owl Nebula http://www.lunarplanner.com/asteroids-dwarfplanets/Sedna.html
116	U.S. Civil War – Battle of Shiloh	http://www.civilwar.org/battlefields/shiloh.html?gclid=CN6c9fPPtKoCFUnc4Aod61cC6w
117	Vitamin K and the Coagulation Cascade	http://www.labtestsonline.org/images/coag_cascade.pdf https://ods.od.nih.gov/factsheets/VitaminK-HealthProfessional/
118	Von-Hippel Lindau Disease	VHL https://my.clevelandclinic.org/health/diseases/6118-von-hippel-lindau-disease-vhl
119	WBC Receptors / Innate Immunity	Toll-like Receptors http://www.sinauer.com/pdf/nsp-immunity-3-10.pdf
120	Who Won the Cold War?	1. China 2. The Deep State
121	Why did the League of Nations Fail?	Look at the United Nations today. Now imagine it without the U.S. Try not to laugh too hard.
122	X-GLUT-5	Flash freeze 500 Joro spiders and mail them to me packed in dry ice.
123	Yuan Dynasty in Medieval China	http://www.thenagain.info/webchron/china/kublaikhan.html

Jared Diamond, *Guns Germs and Steel,* 1997 – The crown jewel of the Jared Diamond collection. Don't leave home without it ☺

Harrison's Textbook of Internal Medicine, 1998 edition – useful as back up when the internet crashes. Can be used as a giant paperweight if all else fails.

Lubert Streyer, Biochemistry (1988) – Not as comprehensive as Voet and Voet but a great text for Biochem 101.

Voet & Voet, *Biochemistry* (2004) – Pricey but worth it, especially for the CD-ROM

Printed by Libri Plureos GmbH in Hamburg, Germany